*Electronic Materials
and Devices*

Electronic Materials
and Devices

David K. Ferry
Arizona State University

Jonathan P. Bird
Arizona State University

ACADEMIC PRESS

A Harcourt Science and Technology Company

San Diego   San Francisco   New York   Boston   London   Sydney   Tokyo

ACADEMIC PRESS
A Harcourt Science and Technology Company
525 B Street, Suite 1900, San Diego, CA 92101-4495, USA
http://www.academicpress.com

Academic Press
Harcourt Place, 32 Jamestown Road, London, NW1 7BY, UK
http://www.academicpress.com

Library of Congress Catalog Number 2001088198

International Standard Book Number 0-12-254161-8

Printed in the United States of America
01 02 03 04 05 06 IP 9 8 7 6 5 4 3 2 1

Contents

CONTENTS

9.5. Giant Magnetoresistance 362
9.6. Magnetic Memory 363
Problems 366

10. *Superconductivity* 369
10.1. Properties of Superconductors 370
10.2. The Meissner Effect 374
10.3. The London Equations 377
10.4. The BCS Theory 380
10.5. Superconducting Tunneling 383
10.6. High-T_c Materials 389
Problems 391

Appendices 393
A. The Hydrogen Atom 393
A.1. Separation of the Angular Equation 395
A.2. The Radial Equation 400
B. Impurity Insertion 405
B.1. Impurity Diffusion 405
B.2. Ion Implantation 408
C. Semiconductor Properties 411
D. Some Fundamental Constants 413

Index 415

Preface

The motivation for this book grew from both of our experiences of teaching an introductory course on the properties of electronic materials to junior and senior students in the Electrical Engineering department at Arizona State University. This course provides students with their first introduction to the basics of quantum mechanics and builds on these principles to provide a quantitative discussion of semiconductor devices (*p-n* junctions, BJTs, MOSFETs, etc.). As an introductory course, it is our opinion that the basic philosophy should be one of keeping mathematical treatments to the reasonable minimum level required to achieve satisfactory rigor while focusing on the key physical concepts that underlie the properties of electronic materials. It is this philosophy that we have attempted to apply in the formulation of this book.

The early chapters are devoted to an introduction of crystal structures and quantum mechanics, and to the application of these basic principles to the problem of electron motion. This discussion leads us to introduce the concepts of band theory in solids which ultimately allows us to appreciate the natural classification of materials into metals, semiconductors, and insulators. Later chapters focus on the application of these different materials in a variety of technologies with a strong emphasis in particular on the applications of semiconductor systems in microelectronics and optoelectronics.

Detailed discussions of dielectric and magnetic materials are also given, however, with the emphasis placed on the application of these in technologies such as MEMS and magnetic data storage. A final chapter focuses on the special properties of superconducting materials, including a brief discussion of high-temperature superconductors. Numerous examples and references to the specialist literature are included throughout the text, and for this reason we hope that this book will also be of use to students in other engineering disciplines, as well as to professionals in the field.

We should remark that we teach the class as a four semester hour class, in which there are four 50 minute lectures each week. The material is far too extensive to do otherwise, although there are options that one could pursue to teach it in a quarterly course. Nevertheless, individual instructors have trouble in getting completely through the material, as a lot of material is presented here.

A number of people provided invaluable assistance in the preparation of this book. We would particularly like to acknowledge all those authors who provided permission to reproduce their work in the book. The pictures of various MEMS structures that appear in Chapter 7 were taken in the group of the late Professor Guckel at the University of Wisconsin who was a prolific contributor to the development of this exciting field of research. We would particularly like to thank his wife for allowing us to reproduce these pictures which we hope will serve as a lasting tribute to the many contributions made by Professor Guckel. Other people deserving of special mention are Sally Kempner at Lucent Technologies and Michelle Evans at Intel. The cover photo is of the potential profile in a small MOSFET and was provided by Richard Akis of Arizona State University. Others who have influenced the material are our colleagues Steve Goodnick, Robert Grondin, Ron Roedel, Dieter Schroder, Trevor Thornton, and Dragica Vasileska.

David K. Ferry
Jonathan P. Bird

CHAPTER 1

Introduction

The explosion of scientific knowledge that occurred in the first 30 years of the 20th century led scientists and engineers to develop a host of new electronic devices. This rapid development was hastened by two major world wars. However, for the first half of the 20th century, the vacuum tube was the primary component of electronic systems. Its operation was well understood from a phenomenological viewpoint and its behavior could be predicted with reasonable success. In other words, the electronics community was a content, happy group, composed of people dedicated to the application of the electron tube to complicated electronic systems. Yet, a multitude of important new concepts were to come together to radically change this stability and to provide for a rapid growth in what we now know as *information technology.*

Michael Faraday[1-3] had already discovered semiconductors in the 19th century. It was the property of the conductivity *increasing* with temperature in AgS, rather than decreasing as in a metal, that identified the material as having new properties. The next important breakthrough was the observation of rectification in a metal–semiconductor contact by Braun,[4] who joined an iron pyrite to PbS. The rectification discovered by Braun proved useful in the new radio electronics, where it could be used for direct detection of the waves,[5,6] even though it was not until 1938 that Schottky[7] provided the theory

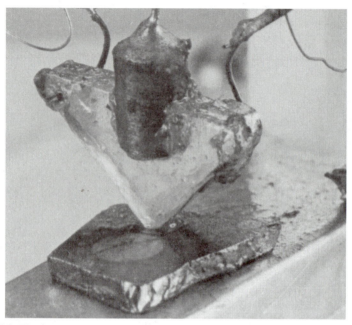

Figure 1.1. The first transistor, as fabricated at AT&T Bell Laboratories. *Copyright © 1997 Lucent Technologies. All rights reserved. Reprinted with permission.*

to explain the action of the metal–semiconductor contact. While it may seem that little was done between 1833 and, say, the Second World War, much work on the properties of semiconductors was carried out, and this led to the discovery of the germanium transistor (Fig. 1.1) by Bardeen, Brattain, and Shockley in 1947.[8,9] The transistor revolutionized the world of electronics, but it was only the beginning.

The idea of constructing a new computing machine grew in several areas and countries,* but the most extensive effort was carried

---

*The idea of an analytical engine had already occurred to Charles Babbage in the mid-19th century. However, it was brought forward by the theoretical concept of Alan Turing.[10] Some initial efforts appear to have begun under the direction of John Atanasoff at Iowa State University (see, e.g., Mackintosh[11]). Like Atanasoff, Konrad Zuse in Germany pursued efforts to build an electronic calculator. Indeed, he went through several generations, and his calculators may have been used in the V2 project near the end of the war (see, e.g., the discussion in Hodges[12]. At the same time, Alan Turing participated in building the Enigma computer for code-breaking applications at Bletchley Park in England during the war.

Figure 1.2. The first integrated circuit, as created by Jack Kilby in 1958. The circuit contains one transistor, three resistors, and one capacitor. *Figure reproduced with kind permission of Texas Instruments.*

out at the University of Pennsylvania, and this culminated in the ENIAC in 1948.** Within a few years, major efforts were underway in several industries and universities. This was coupled to the rapid growth in microelectronics that followed the discovery of the integrated circuit in the late 1950s by Kilby[13] (Fig. 1.2) and that of the metal-oxide–semiconductor transistor.[14–16] (Kilby was finally awarded the Nobel prize for this work in 2000.) These advances finally led to the invention of the microcomputer in 1964.[†] Things have never been the same. There is no obvious end to the growth or extent of the inroads that microelectronics will make in our lives. Indeed, even in everyday objects such as our automobiles, we find multiple microprocessors being used to control the engine, emission

---

** Note that relay-based computers were being pursued at Bell Telephone Laboratories at the same time.

† In general, it seems that the first microprocessor was developed at about the same time by Texas Instruments and Intel. The first largely successful commercial microprocessor seems to have been the Intel 4004, which contained about 2300 transistors.

controls, radio, temperature and climate of the passenger compartment, and even the ride, through an "active" suspension system. Today we live in a "connected" society, where a significant fraction of homes have computers and access the Internet. Information and its control are dominant factors in societal growth and expansion.

One may ask why it is necessary for engineers to understand the physical theory that underlies this continued evolution in information electronics. In fact, there are many who argue that a knowledge of software is adequate. But then, who will provide the new, more advanced systems to utilize better software? Thus, it is imperative that such training be provided. The solid state theory necessary for this treatment encompasses a great deal that is not unique to the theory of solids. Rather, most of the underlying knowledge basic to the understanding of the properties of materials, and of the operation of electronic devices, arises from the field of modern physics. Modern physics itself arose during the first half of the 20th century with breakthroughs in classical mechanics (relativity theory) and the introduction of the new quantum mechanics. The first model of the atom appeared only in 1913.[17] Quantum mechanics burst upon us in 1925–1926 through the work of Heisenberg[18] and Schrödinger[19]. Thus, in the relatively short period of 30 years near the start of the 20th century, physics itself was revolutionized. As we shall see, these theories provide the basis for our understanding of the properties of electronic materials and devices.

From everyday experience, we know that matter may exist in many forms (we discuss these forms later). Moreover, this matter may exhibit a number of different properties such as electrical and thermal conduction, optical transmission, magnetism, and superconductivity. What is not so clear, however, is why certain materials exhibit these different properties. It is clear that some of these properties are explained by the intrinsic electronic states of the materials, but others are not so easily explained. Hence, it is necessary to study just what is known and what is not known in order to utilize materials for given applications. If we could achieve a full, microscopic understanding of materials, it would be possible to predict certain properties. This, in turn, would enable us to synthesize materials with a given set of properties. To succeed at this, it is necessary to understand the basis of the properties from the atomic and molecular level, and to understand how small differences in structure can lead to large differences in observed properties. Our goal here is to achieve this microscopic understanding of materials properties, at least to the level that this can be done. First, however, let us examine one of the most important driving forces for new materials developments—modern very large scale integration (VLSI) of electronic circuits.

1.1. MODERN VLSI

As the size of the transistor was made smaller, and transistors began to replace vacuum tubes, the size of the computer became considerably smaller. Not the least of the driving forces for this to occur was the considerably lower power dissipation in the transistor, relative to that in the vacuum tube. This reached a new plateau with the invention of the microprocessor. The microprocessor took the heart of the computer—the central processing unit—and placed it, together with a small amount of cache memory, on a single integrated circuit chip. While the first such microprocessors could do only simple calculations, their capabilities have essentially doubled every 1.5 years since these first ones. As a result, today's microprocessors are essentially well beyond what was considered a supercomputer just a decade ago. The modern microprocessor is capable of carrying out more than a billion instructions per second. Such a chip is shown in Fig. 1.3(a), where we show the view of the Pentium III chip, manufactured by Intel, Inc. By itself, one does not get an impression of its overall size, so the packaged chip is shown on a keyboard next to a postage stamp in Fig. 1.3(b).

The initial transistors were simple structures. Today, the heart of the microprocessor is the silicon metal-oxide–semiconductor (MOS) field-effect transistor (FET), developed shortly after the first transistor.[14–16] The coupling of n-channel and p-channel devices into a single complementary MOS structure created the low-power breakthrough that enables us to put millions of transistors upon a single integrated circuit without the entire structure melting from the power dissipation (we discuss these devices in Chapter 6). One such single transistor is shown in a high-resolution cross-sectional electron micrograph in Fig. 1.4.

(a)

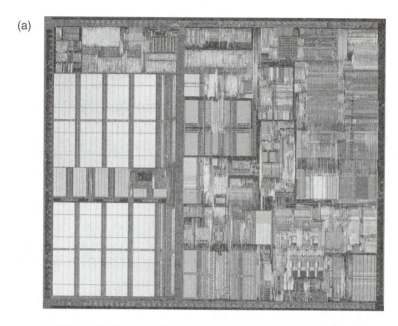

(b)

Figure 1.3. (a) A view of the Pentium III microprocessor, which is an example of the modern-day integrated circuit. This chip contains more than five million transistors. (b) The packaged Pentium III chip is shown next to a postage stamp to give an indication of its small size. *Images reprinted by permission of Intel Corporation. Copyright © Intel Corporation 2000.*

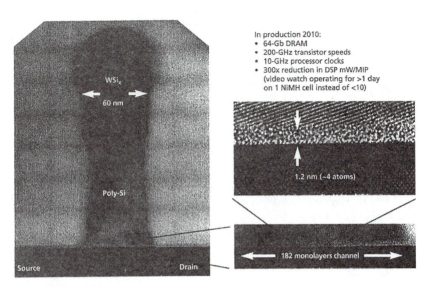

In production 2010:
- 64-Gb DRAM
- 200-GHz transistor speeds
- 10-GHz processor clocks
- 300x reduction in DSP mW/MIP
  (video watch operating for >1 day
  on 1 NiMH cell instead of <10)

Figure 1.4. A cross-sectional image of a 60-nm MOSFET built at Bell Labs. The atomic lattice is visible in the highest magnification view of the 1.2-nm gate oxide. *Copyright © 1997 Lucent Technologies. All rights reserved. Reprinted with permission.*

It may seem that this growth has relied almost exclusively on the properties of semiconductors, but this is not the case. A modern microprocessor depends on semiconductors, metals, and insulators for different usages within the circuit. Each must be prepared through the manufacturing process to have specific properties and to be localized to specific areas of the chip under the most stringent manufacturing tolerances. If we think about the important length in a transistor (as we will see later, this is the gate length), the various different materials must be aligned to essentially one-third of this value. If the gate length is only 0.18 micron (1 micron, or μm, is 10^{-6} m)[‡], then the alignment must be accurate to 0.06 μm, or 60 nanometers (1 nm = 10^{-9} m). In silicon, the distance between atoms is only about 0.24 nm, so this means that manufacturing must produce alignment to within about 250 atoms displacement! The

---

[‡]It should be recognized that the modern semiconductor industry leads a paranoid existence in terms of dimensions. The thickness of the chip is measured in mils (1 mil = 10^{-3} in.), an English unit, whereas the lateral dimension is measured in metric units, microns for devices and millimeters for wafers..

semiconductor is the host material, but the oxide is the crucial insulator in the transistor, and the metal provides interconnects to distribute power and signals across the chip. It is easy to understand that the science of manufacturing has become one of the most important disciplines in the high-technology industry today. Moreover, it must be recognized that progress is achieved by continuing to reduce each of these dimensions.

1.2. THE DRIVING FORCES FOR CONTINUED INTEGRATION GROWTH

What is the reason that we mark continued progress by the reduction of the dimensions for each succeeding generation of integrated circuit? Quite simply, it is *economics*! As manufacturing technology continues to evolve, the process by which integrated circuits (and, indeed, nearly any other high-technology product) are fabricated becomes more efficient and cost effective. In the semiconductor industry, this has meant that the cost of processed silicon chips, fully tested and packaged, remains roughly at $10–20$ cm^{-2}. This cost has been almost constant for the past 4–5 decades, yet the complexity of the circuitry on each square centimeter has increased by orders of magnitude.

In economic terms, the driving force is the cost per function that is implemented. In early computers, a multiplier circuit consumed many chips and was very expensive to implement. As a result, only the top-end scientific computers were so equipped. Today, however, the hardware multiplier is an integral part of almost all microprocessors. Consequently, the cost to implement this circuitry is now measured in fractions of a dollar. This trend has continued as more functions are implemented upon the single-chip processor. Such a trend has led to the capabilities that are now utilized in single-chip controllers for a variety of products. Today, these are often referred to as "smart" products, and future trends will have a number of routine household appliances interconnected either through a local area network or through the internet so that a central monitoring computer can monitor the performance and "health" of the appliances.

The growth in functionality is also inherent in the microprocessor itself, and today's implementations are far more capable as computers than supercomputers of the past decade. As the dimensions of the individual transistors, and other components, shrinks from one generation to the next, it is possible to put more functions on the same chip area, hence reducing the cost of each of the functions. This cost reduction drives the continued evolution of the chip itself. Of course, there is a scaling rule that has governed this evolution, and this is known as "Moore's law," after Gordon Moore who was one of the founders of Intel.

1.3. MOORE'S LAW

Gordon Moore summarized the growth of the integrated circuit over its existence in 1974 in a paper at the annual Electron Device Meeting.[20] He pointed out that, from its inception, the integrated circuit had grown at a rate for which the number of transistors on each chip quadrupled each generation—about every 3 years. This growth curve is not unique to integrated circuits, and has appeared in many other fields, but here it represents the constant advance in technology driven by a continued increase in the number of functions included on each chip, as already discussed. This growth has come to be called *Moore's law*. Moore himself pointed out that the growth actually had three distinct, and important, components.

The first of these components is the reduction of device dimensions, which has already been discussed. Each generation sees a reduction of the dimension by about a factor of $\alpha = 1/\sqrt{2}$. This, by itself, gives an increase in density of a factor of 2 each generation. We will see in Chapter 6 what this entails for the scaling of the MOS transistor. The second component of growth is an increased circuit complexity, or cleverness. That is, novel changes in the circuit itself lead to effective increases in density. One example of this is the introduction of the trench capacitor for memory cells. Quite often, the dynamic memory is implemented merely by storing charge on a capacitor on the chip. If the capacitor is charged, the state is a binary "1," whereas if the capacitor is uncharged, the state is a binary "0." In the early days of such memories, the capacitor was a normal metal-oxide–semiconductor structure, such as that shown earlier in Fig. 1.4. It was discovered, however, that one could save space on the chip by cutting a trench deep into the chip, growing an oxide on the sides of the trench, and then filling the latter with a metal. This turned the capacitor from its planar configuration to a vertical one, and saved an enormous amount of space on the chip, which enabled a much greater packing density to be achieved. This is just one such "clever" circuit change that allows for an increase in the density on the chip.

The third factor, according to Moore, is an increase in the size of each chip. If the density is increased by the first two factors, then why should the chip increase in size? The answer is the ever-prevalent economics: more functions on the chip mean a lower cost per function. This factor represents, in fact, the increase in manufacturing capability, but the increase in chip size has other important consequences. Consider for example the semiconductor wafer upon which the individual chips are fabricated. The wafer is round, but the chips are square or rectangular. The layout of rectangular chips on a round wafer means that chips at the edge are not complete (parts of the chips are off the wafer), and become waste material, which can be seen in Fig. 1.5, which shows a wafer of Pentium III chips. Economics again suggests that one needs to make a minimum number of chips per wafer; so an increase of chip size means an increase of wafer size.

Indeed, new fabrication facilities (called "fabs") brought on line around 1997–1998 for the 0.25-μm technology used 200-mm wafers. Fabs that are planned for 2002+, some two generations down the road with 0.13-μm technology, will utilize 300-mm wafers. It is easy to extrapolate to the not too distant future where wafers will be the size of garbage can lids! It is also easy to understand the economics. If a 200-mm wafer can contain some 150 microprocessors, each of which can be sold for more than $650, this wafer is worth some $100,000 when the packaged chips are finished. Moore's law is now seen as a representation of the economic driving forces in this industry. It is no more, nor any less, than the reflection of the need for continued growth and the factors that must fall into place to enable this growth. Hence, the reduction in dimensions is not a convenience, but is required to meet the continued pressure of the economic driving forces themselves.

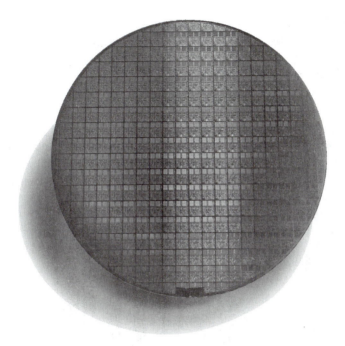

Figure 1.5. A wafer of Pentium III chips. Near the edges of the wafer, individual chips are incomplete. and this accounts for the lost area of the wafer. *Reprinted by permission of Intel Corporation. Copyright © Intel Corporation 2000.*

But, this digresses from our major task. To make integrated circuits and/or microprocessors, we need to understand and control materials properties. The purpose of this book is, in fact, to discuss just these properties, and how they differ from one material to another. To prepare for the chapters ahead, we now turn to the various categories of materials that will be important, and how this book is laid out to discuss the pertinent properties.

1.4. TYPES OF MATERIALS

In classifying matter, one normally thinks about the states of matter: solid, liquid, or gas. Of course, here we think of electronic materials as solids. Yet, there are some materials, which appear to be solids but possess some liquidlike properties — the liquid crystals. How do we differentiate between these different phases? The answer is that it must be done quite carefully, and with some idea of the applications and properties that are under consideration.

In gases, the interaction between the individual atoms or molecules is quite weak, so that the motion is largely uncorrelated. These individual molecules move in an apparently random manner and have a broad distribution of velocities. In liquids, the density of molecules is much larger, but the molecular interactions are still relatively weak so that there is no structural stability. When we get to solids, the density of atoms or molecules is not significantly higher than that of liquids, yet the solids tend to come together with a crystal structure, which entails some mechanical stability.

While we have asserted that the solid has some crystal structure, this is not always the case. The simplest solid is an *amorphous* solid, where there is no underlying crystal structure. However, we generally think of amorphous materials as being special cases of crystalline materials in which the underlying crystalline structure has been destroyed by some means. Amorphous materials often have the same short-range order as crystalline materials, but do not possess the same long-range order. Between these two extremes are the polycrystalline materials. In Fig. 1.6, a high-resolution electron micrograph shows the existence of a single grain of crystalline silicon embedded in a matrix of amorphous silicon. Polycrystalline material is composed of many such grains — indeed, it may well be the case that all

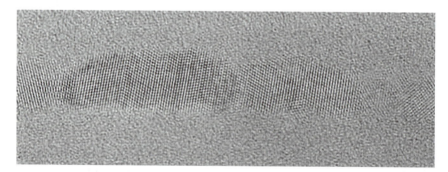

Figure 1.6. Transmission electron micrograph showing a crystalline grain of silicon that is embedded in a matrix of amorphous silicon. *J. Lützen et al., J. Vac. Sci. Technol.* **16**, *2802 (1998). Figure reproduced with the permission of the American Institute of Physics.*

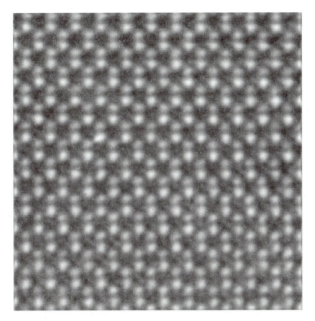

Figure 1.7. Lattice image of the zinc-blende crystal structure of GaAs. Individual atoms are clearly resolved in this image. *Picture provided courtesy of A. Trampert and O. Brandt.*

of the material is crystalline, but belonging to a great many different grains. The orientation of the crystallites is random in these polycrystalline materials.

The crystalline materials possess both short-range and long-range order. They have a very specific crystal structure (to be discussed in the next chapter). For example, silicon possesses the diamond structure, in which the basic unit is a face-centered cube with two atoms at each corner and face of the cube. In Fig. 1.7, the crystalline structure of a zinc-blende lattice (GaAs) is illustrated by a high-resolution lattice-plane image produced by electron microscopy. We will return to this picture in the next chapter and explain what is seen in the image.

Then, one must characterize the electronic properties of the material. Solids can be metals, insulators, or semiconductors (which are poor insulators, but also poor metals, and for which the properties can be adjusted by the introduction of impurities in a controlled manner). They can be good conductors or poor conductors, and they

Figure 1.8. Magnetic domains are imaged by phase contrast microscopy. The solid lines are phase contrast fringes, while the light and dark regions are different magnetic domains. *Picture reprinted with permission of Lucent Technologies Bell Labs.*

may exhibit other properties such as magnetic domains or superconductivity. A small magnetic domain is illustrated in Fig. 1.8. All of these different properties must be described together in order to fully characterize any individual material.

Semiconductors, which are the important material for transistors and integrated circuits, are predominantly from the group IVB column of the periodic table. This column has carbon (which produces diamond), silicon, germanium, and tin. These materials have four valence electrons and form tetrahedrally coordinated, covalent bonds. Compound semiconductors come from columns IIIB and VB, so that they average four valence electrons, or columns IIB and VIB. It is the tetrahedral coordination and covalent bonding that gives the semiconductors their unique properties that have become so useful for modern electronics.

Yet, other materials such as the metals remain important. The dielectric and magnetic properties provide new opportunities for modern electronics, so that we cannot merely focus on the semiconductors. Yes, they are important today, but the future may well require a different set of properties and a different set of materials to continue the growth of technology. Hence, we must study all of these properties.

We begin this in the next chapter, where we study the crystalline nature of the materials. We then turn to the electronic properties, which are described through their quantum mechanical nature. Then, a discussion of the importance of semiconductors, and how one tailors them to achieve desired properties leads us into the conduction properties of electrons (and holes). This leads us naturally to a discussion of electron devices. The properties of dielectrics and magnetic materials then follow before we close with a discussion of superconductivity.

REFERENCES

1. M. Faraday, *Experimental Researches in Electricity*, Ser. IV, pp. 433–439 (1833).
2. M. Faraday, Beibl. Ann. Phys. **31**, 75 (1834).
3. T. Martin, *Faraday's Diary*, Vol. 2 (London: G. Bell and Sons, Ltd., 1932), pp. 55–56.
4. F. Braun, Ann. Phys. Pogg. **153**, 556 (1874).
5. J. C. Bose, 1904, U.S. Patent No. 755840.
6. G. W. Pierce, Phys. Rev. **25**, 31 (1907).
7. W. Schottky, Naturwissenschaft **26**, 843 (1938).
8. J. Bardeen and W. Brattain, Phys. Rev. **74**, 230 (1948).
9. W. Shockley, Bell Syst. Tech. J. **28**, 435 (1949).
10. A. Turing, Proc. London Math. Soc., Series II, **42**, 230 (1936).
11. A. P. Mackintosh, Sci. Am. **259**(2), 90 (August 1988).
12. A. Hodges, *Alan Turing: The Enigma* (New York: Touchstone, 1983), 299.
13. J. Kilby, IEEE Trans. Electron Devices **23**, 648 (1976).
14. W. Shockley and G. L. Pearson, Phys. Rev. **74**, 232 (1948).
15. J. L. Moll, IEEE Wescon Conv. Rec., Pt. 3, 32 (1948).
16. W. G. Pfann and G. C. B. Garrett, Proc. IRE **47**, 2011 (1959).
17. N. Bohr, Phil. Mag. **26**, 476 (1913).
18. W. Heisenberg, Z. Physik **33**, 879 (1925).
19. E. Schrödinger, Ann. d. Physik **79**, 361, 489 (1926); **81**, 109 (1926).
20. G. Moore, *IEDM Techn. Dig.* (New York: IEEE Press, 1975), 11.

CHAPTER 2

The Crystalline Nature of Materials

In the development of electronic materials and devices, the structure that we call the solid state is of primary interest. That is, materials that appear in a condensed state are our major focus. These materials must, in addition, possess an ordered, or crystalline, structure in their atomic arrangement, which we discuss in this chapter. It is this periodic crystalline structure that provides most of the properties of interest in electronics. For example, it is this regular atomic structure that provides the basis of the band model of solids and leads to electrical conduction. Moreover, it is the disruption of the regular atomic structure at the edge of a crystal that leads to the surface effects of interest, such as surface potential barriers. In considering the origin of this crystalline structure, we begin by studying the nature of atomic bonding.

There are several forces that act on a collection of atoms or molecules. These forces, both attractive and repulsive, are dependent on the relative distances between the various atoms. In crystal systems in which electrons are transferred among atoms there is, of course, the relatively long range Coulomb-type force, which is primarily attractive in nature. This force arises from the exchange of electrons from one atom to its neighbor, and is a type of ionic attraction. Ionic attraction is just one of the possible forces acting on the various atoms. We shall talk about this and other forces presently.

One might expect that the attractive forces acting on the atoms would pull them together until they collide. However, there exists a very short range repulsive force acting between the nuclei and electrons of the individual atoms. As a result, there is some equilibrium distance at which these forces exactly balance each other. At this equilibrium distance, the attractive force—be it Coulombic or some other force—is exactly counterbalanced by the repulsive force between the two nuclei and between the two electron distributions. If the atoms move apart, then the repulsive force is smaller than the attractive force, and they are pulled back together. If they move closer together than the equilibrium distance, the repulsive force is the stronger one, and they are pushed apart. Because the forces tend to always make the atoms move back to the equilibrium position, this position is a stable equilibrium. We can think of these forces as arising from potential energies (we recall that the force is related to the potential as $F_x = -\partial V/\partial x$ in one dimension). Since the equilibrium is stable, it must occur at the potential energy minimum.

Figure 2.1 helps clarify these statements. Here, F_A, F_R, and F_T are the attractive, repulsive, and total forces, respectively, that act on the atoms. The attractive force is the aforementioned Coulomb force, while the repulsive force is the short-range force that keeps the electron from penetrating the atomic core. Note that there is a point at which $F_T = 0$; that is, the total force is equal to zero—the attractive force and the repulsive force exactly cancel one another. This is the equilibrium atomic spacing a_0. The ensemble of atoms all reside in a series of potential wells, described by the totality of forces between the individual atoms. Given a small perturbation about the equilibrium position a_0, the atoms tend to return to that position. Thus, as already noted, the system is in a stable equilibrium. The atoms may still vibrate about the equilibrium position due to agitation from their thermal energy.

2.1. THE VARIOUS STATES OF MATTER

From Fig. 2.1, two parameters can be discerned that will enable us to identify certain properties of solids. The first of these is the

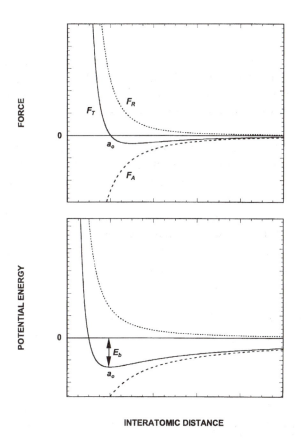

Figure 2.1. Relationship of atomic forces and potential energy. The equilibrium position is that at which the net force is equal to zero and for which the corresponding potential energy exhibits its minimum.

equilibrium interatomic distance, or lattice constant a_0. This is the equilibrium spacing already discussed. The second quantity is the binding energy E_b. As the name implies, the binding energy is just the energy per atom required to dissociate, or break up, the solid. This energy is, of course, a measure of the strength of the solid, and varies from as little as 0.02 eV* for He_2 (van der Waals bonding), up to

*The symbol eV refers to electron volts. One eV is the energy gained by accelerating one electron through one volt, and $1 eV = 1.6 \times 10^{19}$ Joule — the conversion from Joule to eV is made by dividing the former by the charge on the electron.

more than 10 eV for LiF (ionic bonding). The binding energy is shown in Fig. 2.1 as the minimum of the total potential energy curve.

As the binding energy is reduced in magnitude and the lattice constant a_0 increases, the atoms move farther apart. As this process continues, the binding becomes so weak that the atoms are no longer held in a lattice array and are free to move. In this case, it is said that the group of atoms is a *liquid*. For further increases of a_0, the atoms cease to interact at all, and they are essentially free atoms. This is the case for a *gas*.

As evidenced, the various types of bonding that occur in nature can be distinguised by differing degrees of strength. It is pertinent at this point to discuss these types of bonding, although we will return to this discussion in a later chapter where we discuss the quantum mechanical nature of the bonding. Of course, from chemistry, ionic and covalent bonding are already familiar. Note (from the following descriptions) that the bonding classification of solids is primarily described by the dominant attractive force. Let us now take a closer look at these various types of bonding.

2.1.1. Van der Waals Bonding

When two unlike electrical charges are separated a small distance they are said to form a dipole, and to have an associated dipole moment. (We shall discuss these dipole moments further in Chapter 7.) For now, suffice it to say that forces are produced which may interact with other charges or dipoles, and we depict this schematically in Fig. 2.2. When the electron distribution of an atom is shifted in position slightly with respect to the nucleus, a dipole moment is established. The dipole moments of a large number of such atoms, in which the electrons have been shifted relative to the ions, may interact in such a way as to cause the atoms to be attracted to each other. The shift in the electronic charge, relative to the ionic charge, gives the dipole itself. Then, one dipole can interact with another, as one electronic charge distribution is attracted to the opposite ionic charge. The two electronic charges repel, as do the two ionic charges. Yet, a net attractive force can occur under certain situations.

In general, these dipole interaction forces are very small, and van der Waals bonding is not of primary importance to a discussion of

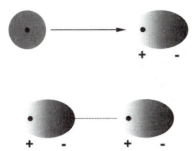

Figure 2.2. Spontaneous dipole formation and van der Walls bonding. The upper figure shows that although an atom is neutral overall, an atomic dipole may exist at any point in time due to the motion of electrons in the atom. The lower figure illustrates the formation of a van der Waals bond (dotted line) between two spontaneously formed dipoles.

the materials of interest here. However, it is important to note that there are two types of van der Waals bonding, in that the dipoles can either be induced or permanent. For induced dipoles, we think about materials such as argon (A) in which the dipole doesn't exist until the solid begins to form. Then, the dipole arises as a response to the forces between the two nucleui, the two electron clouds, and one nucleus with the opposite electron cloud. On the other hand, water molecules possess a permanent dipole, and the formation of ice is the interaction between these dipoles. Although we shall talk about dipole forces in a later section, we shall not discuss van der Waals bonding further. Nevertheless, the dependence of the potentials on energy have given rise to a standard description that is often used for a variety of other bonding types. It is often said that the attractive forces vary as r^{-n} while the repulsive forces vary as r^{-m}, with $m > n$ required if the bond is to be stable. A common potential is the 6–12 potential ($n = 6$, $m = 12$), for which the potential energy of Fig. 2.1 can be written as

$$E(r) = \frac{A}{r^{12}} - \frac{B}{r^6},$$ (2.1)

and for which the equilibrium distance is found to be

$$a_0 = \left(\frac{2A}{B}\right)^{1/6}.$$ (2.2)

We will return to this form later.

2.1.2. Ionic Bonding

In ionic bonding, one of the atomic constituents yields one or more outer shell electron(s) to the other atom, producing a set of positive and negative ions that attract by means of a Coulomb attraction. For example, in a NaCl molecule, a single electron is transferred from Na to Cl, and we obtain the following expression for the attractive force:

$$F_A = \frac{e^2}{4\pi\varepsilon r^2}. \tag{2.3}$$

Here, ε is the dielectric permittivity of the medium in which the charges reside, and r is the distance between the two charges. In the equilibrium position, we have $r \to a_0$. Thus, the binding energy is just the electrostatic potential energy evaluated at the equilibrium position. This equation, however, is for two charges. In a solid crystal, the forces arise between all the nearest-neighbor atoms and the central atom, so that the sum of forces on any single atom must balance for equilibrium.

2.1.3. Covalent Bonding

In covalent bonding, electrons are shared between two or more neighboring atoms. For example, the atomic form of hydrogen can accommodate two electrons in its outer shell, but the bare atom has only one. Therefore, two hydrogen atoms can form a molecule in which they share their two electrons between themselves. This type of bonding is found in most organic compounds and in nearly all semiconductors. In germanium and silicon, which each have four valence electrons (eight are possible in this shell), each atom is held to its four nearest neighbors by covalent bonds. The four electrons from each atom are shared with its four neighbors, so that the bond between any two atoms has (on average) two electrons. These bonds form a tetrahedron, and are highly directional in nature. This accounts for some of the observed crystal properties of such materials. Again, we shall return to this in Chapter 4.

2.1.4. Metallic Bond

This type of bonding is very similar to covalent bonding, except that the valence electrons are shared by a great many atoms. In metals, the valence electrons are essentially free electrons and move throughout the metal; hence they are not restricted to the strong bond position of the covalent bond. As a result, they are shared by all the atoms in the metal. This type of bond is not directional, and the free electrons account for the high conductivity of metals.

2.2. SPACE LATTICES

So far, it is apparent that solids are held together by their binding forces. In addition, they are crystalline in nature. By crystalline, we mean that the atoms are arranged in a regular, periodic array. There are, however, some special cases of this idea. One can have amorphous solids, in which the crystalline nature is very short range and the crystal has no long range order. On the other hand, a polycrystalline solid has many *grains*, each of which is single crystal material, but which are randomly oriented with respect to one another. Indeed, Fig. 1.6 showed a single polycrystalline grain embedded within an amorphous crystal. Finally, one can have *single crystals* in which the long-range order of the lattice extends throughout the crystal. These ideas are shown in Fig. 2.3.

In a single crystalline solid, the interatomic distance a_0 is the distance between the two closest atoms in a particular direction. Since solids are generally three-dimensional, it is possible that a different lattice constant a_0 occurs for each direction. This raises an interesting point—it is possible that for some solids there is no crystalline lattice order, as discussed earlier where these materials were called *amorphous* materials. Strictly speaking, amorphous materials are crystalline solids in which the lattice has become disordered. As a result, there is no long-range order in the solid, even though the short-range structure is quite like that of the crystalline solid. In the crystalline version of the material, the order is assumed to be *perfect*. That is, every atom is in its predetermined position in the crystal structure. But, what is this structure?

Figure 2.3. Different types of order in crystalline solids. The left picture illustrates an amorphous arrangement of atoms in which there is little evidence for long-range order. The center picture illustrates a polycrystalline material in which isolated crystallites with random relative orientations are embedded in an amorphous matrix. Finally, the right picture illustrates the ordering of atoms in a perfect crystal structure.

Figure 2.4 shows part of a lattice array, in which there are two distinct distances, a_0 and b_0, which are the *lattice constants* or interatomic distances in the two different directions within the crystal. (Note that these distances are not equal. The reason for this is not important for our purposes; suffice it to say that it is due to microscopic binding mechanisms.) However, it is also possible for each lattice site to possess a *basis* (shown in the lower set of Fig. 2.4), which is an arrangement of atoms around the lattice site. Thus, the crystal structure is defined by both the underlying lattice and the basis at each lattice site. We will explore this with some examples in the following discussion. Also note that we only show two dimensions in Fig. 2.4. Real crystals are three-dimensional, and the third dimension is normal to the page of Fig. 2.4. We will return to this point later. Whether two-dimensional or three-dimensional, there is the *lattice* itself, and, second, there is a *basis* attached to each lattice point.

If we choose just four of the atoms, such as four of the lattice sites in Fig. 2.4, and place them end to end and side to side, it is possible to reproduce the entire lattice. Such a cell, which may be replicated throughout the lattice, is called a *unit cell*. To study the entire crystal structure, we need only study the unit cell, since the crystal is made up of replications of this basic structure. Before embarking on a study of the unit cell, however, let us formally define a lattice: A *lattice* is

LATTICE   BASIS   CRYSTAL

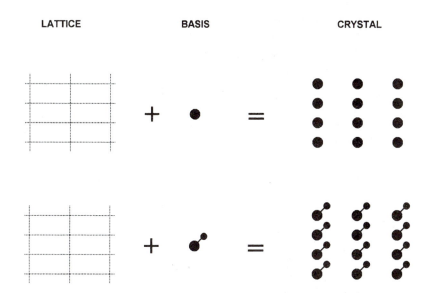

Figure 2.4. Crystal structures may be broken down into a *lattice* and a *basis*. In the examples shown here, the lattice points correspond to the vertices of the dotted lines. Both crystal structures shown exhibit this *same* lattice structure but differ in the details of their basis units.

simply a periodic array of points, generally in three dimensions, usually representing the positions occupied by the atoms of the crystal. Each point of the lattice may be represented by a vector, which is drawn from the origin to the lattice point. The important aspect of this is the existence of minimum vectors for the unit cell.

Figure 2.5 shows a general two-dimensional lattice and a set of vectors which define the various lattice points. In three dimensions, the unit cell is characterized by its three primary directional lengths a_0, b_0, and c_0 and by the three angles α, β, and γ between the three directions. In general, the *triclinic* lattice (Fig. 2.6) possesses no symmetry other than translational symmetry (translational symmetry means merely that the lattice can be reproduced by setting the cells side to side and end to end). However, most of the materials of interest to us possess much more symmetry than this basic lattice. In fact, many of the materials of interest in electronics are those crystals which possess a cubic unit cell; that is, $a_0 = b_0 = c_0$, $\alpha = \beta = \gamma = 90°$. The cubic cell lattice is just one of seven possible lattices and is shown

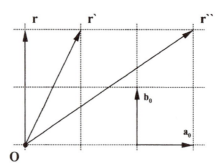

Figure 2.5. *Translation vectors* are used to map between different points in a crystal lattice. In the example shown here, the vectors $\mathbf{a}_0$ and $\mathbf{b}_0$ are referred to as *primitive vectors* because they allow for the mapping between all points within the crystal. For example, $\mathbf{r} = 2\mathbf{b}_0$, $\mathbf{r}'$ $= \mathbf{a}_0 + 2\mathbf{b}_0$, and $\mathbf{r}'' = 3\mathbf{a}_0 + 2\mathbf{b}_0$.

in Fig. 2.7. Materials such as manganese (Mn) and nitrogen (N) possess the simple cubic lattice. In general, however, very few solids possess this structure.

It is perhaps unfortunate that all in life is not so simple. If the simple cubic was all that was needed, the study of crystals would be much easier. There are, however, several, more complex cubic systems which are of interest to us. The first of these is the *body-centered cubic* (BCC). In this structure, an additional atom, or lattice point, is

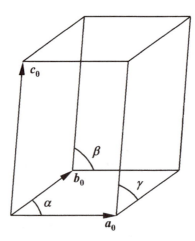

Figure 2.6. Unit cell, translation vectors, and unit-cell angles for an arbitrary *triclinic* lattice.

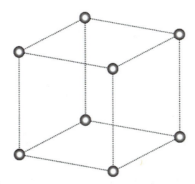

Figure 2.7. Unit cell of the *simple-cubic* lattice.

placed at the center of each cubic unit cell. In the case of NaCl, the basic cell is the BCC with a basis of one Na and one Cl atom per lattice point. The BCC lattice is shown in Fig. 2.8. In some sense, this crystal is composed of two interlocking simple-cubic lattices, in which the second is displaced along each axis by one-half of the lattice constant. Now, one might naively think that we could describe the BCC lattice in just this way: two interlocking cubic lattices. But then, we would have two lattices, not one. It is important that we have only one lattice, which is an important convention. On the other hand, why not describe the BCC as a simple-cubic lattice with a basis of

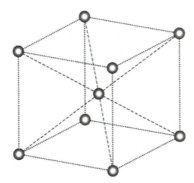

Figure 2.8. Unit cell of the *body-centered-cubic* lattice. This lattice structure is similar to the simple-cubic form already discussed, although an additional lattice point is now present at the center of the cubic unit cell.

two atoms per lattice site, one at the cube corner and one displace by $a_0/2$ in each of the three directions? This would seem to be a valid description, but we discover later that this lattice would have different symmetry properties than are observed from the BCC. For example, if we take a plane diagonally through the crystal, the central atom lies in this plane, so that the lattice distance between atoms (on the lattice site) is different from the same plane passed through the simple cubic structure. This will show up in the x-ray diffraction discussed later in the chapter. As a result, we can draw an important conclusion regarding the basis: the basis of two (or more) atoms arises when one atom is placed at the lattice point and the other atom(s) does not lie on a normal lattice point. To be sure, for the BCC, if we were to assign a basis and use the simple cubic, the second atom would not lie on a point of the simple-cubic lattice. But, we can use the BCC to describe the structure, and use its higher symmetry properties as a result. In general, a basis is used when it is impossible to find a higher symmetry lattice to describe the system. Thus, the convention is to use the highest possible symmetry lattice to describe the system, and only then resort to the use of a basis in the lattice structure.

The second important lattice is the *face-centered-cubic* (FCC) lattice. In the body-centered-cubic lattice, there is an atom located at the center of the cube as well as at the eight corners. In the face-centered-cubic lattice, an additional atom is located at the center of each of the six faces of the cube (see Fig. 2.9).

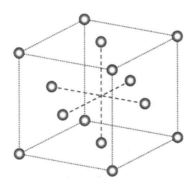

Figure 2.9. Unit cell of the *face-centered-cubic* lattice. This lattice structure is also related to the simple-cubic lattice, but its primitive cell contains additional lattice points at the center of each of the six faces of the cubic unit cell.

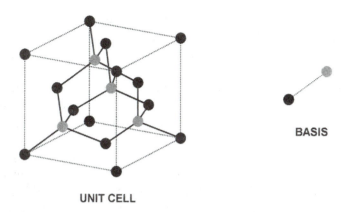

UNIT CELL   BASIS

Figure 2.10. The primitive unit cell of diamond. The lattice structure of diamond is face-centered cubic. The basis consists of two atoms, which are indicated as black and gray balls (for the sake of clarity, atoms that lie outside of the unit cell are not indicated in this figure). The tetrahedral chemical bonding between different atoms is indicated by the solid lines.

This discussion refers to the development of the *space lattice*. It is possible that at each lattice site a *basis* may appear, as discussed. We must remember that the basis is just an ensemble of atoms collected about each lattice point, when we cannot use a higher symmetry lattice to describe the structure. Such crystals are called *lattices with bases*. In this case, the basic lattice is the *space lattice* and the *basis* is the complex of atoms located at each lattice site. For example, in germanium and silicon, the lattice is the face-centered cubic, with a basis of two atoms per lattice site. For the lattice site at the corner of the cube, one atom lies at the corner and the second lies on the body diagonal, one-quarter of its length away from the corner (see Fig. 2.10). This crystal structure is called the *diamond structure*, and is shared most commonly by diamond, germanium, silicon, and gray tin.

In keeping with the preceding discussion, we note that it is impossible to use a higher-symmetry crystal lattice to describe the structure. If we tried to use, for example, the lower one-eighth of the FCC lattice as a primitive BCC, we lack atoms at four nonadjacent corners, so this is not a possible description. Indeed, the FCC is one of the highest symmetry lattices possible, and we adopt the description with a basis for diamond (and Si and Ge) because there is no alternative.

If the two atoms per lattice site in the diamond structure (the atoms of the basis) are different, such as they are in the compound semiconductors, then the lattice is called a *zinc-blende* lattice. Some of the more common semiconductors with the zinc-blende lattice are the III–V compounds InSb, InAs, GaAs, and GaP, and the II–VI compounds ZnTe, HgTe, and CdTe. In indium antimonide, for instance, the basis is composed of one indium atom and one antimony atom.

There is another form closely related to the diamond lattice, which is a *relaxation* of the structure. In general, the diamond lattice is not the equilibrium structure; rather it is metastable (which means that it exists in an *unstable* equilibrium condition—if it is not disturbed it will stay in this structure). The element carbon more often occurs as graphite than it does as diamond. The graphite structure is a relaxation of the diamond structure into a hexagonal planar lattice with weak vertical bonds between adjacent planes. Hence there is strength within the plane, but the planes may move easily with respect to each other. This is what gives graphite its good lubrication properties—the planes slide easily with respect to one another. Similarly, the zinc-blende structure for compound semiconductors is metastable. The relaxation is to a wurtzite crystal structure, which is a variant of the hexagonal structure. Here, the planes do not move so easily, but the structure clearly is different from the zinc-blende and diamond structure of the cubic semiconductors. The wurtzite structure is shown in Fig. 2.11.

An important aspect of the diamond and graphite lattices (and, of course, of the zinc-blende and wurtzite lattices) is that the atoms are tetrahedrally coordinated. Each atom has four nearest neighbors, which sit at the corners of a regular tetrahedron (the reference atom is located in the center of the tetrahedron). This arrangement is easily seen in Figs. 2.10 and 2.11. The difference in Fig. 2.11 is a strengthening of the bonds in the hexagonal plane and a weakening of the bond in the vertical direction. This arrangement is a common attribute of covalently bonded materials, as the covalent bonds themselves are highly directional. In fact, most of the materials in which we will be interested, especially for semiconductor electronics, are tetrahedrally coordinated. Consequently, they have either a diamond, zinc-blende, or wurtzite lattice.

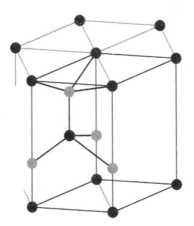

Figure 2.11. The wurtzite crystal structure. The hexagonal unit cell of this structure is denoted by the solid lines, whereas in the upper layer, the full hexagonal structure is indicated for clarity. Bold solid lines are used to indicate tetrahedral bonding between atoms.

2.3. CRYSTALLINE DIRECTIONS

In studying the crystalline structure of solids, we need some standard for defining various directions within the crystal. Obviously, there are a great many ways in which this can be done. One approach, however, has become well accepted and we will follow through with this. Here, we define a coordinate system with respect to the unit cell. That is, for a cubic lattice, the axes of the normal *x*, *y*, and *z* directions are aligned with the edges of the unit cell. This, of course, works best when this unit cell is one of the cubic structures, for the axes then can be interpreted as the normal Cartesian coordinates with angles of $\pi/2$ between any two of the axes. If we were to use a hexagonal lattice, however, then we would need three axes within the hexagonal plane and a fourth axis normal to this plane. Such a complication is beyond our present level of understanding, so we shall stick to the cubic crystal structures.

Directions within the crystal are then easily related to this coordinate system. Figure 2.12 shows this for the cubic cell. Note that the origin of the coordinate system is chosen to be at the back lower corner of the cube. This is entirely arbitrary, since the cube is

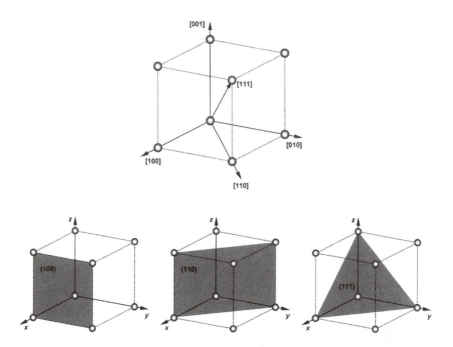

Figure 2.12. The upper figure shows different directions in the cubic unit cell and the corresponding Miller-index notation that is used to describe these directions. The lower figures show different crystal planes in the cubic crystal and the corresponding Miller indices for these planes.

symmetrical. If the cube were rotated, the origin would be at a different corner, but because of symmetry it would not be possible to know the difference. Likewise, all faces of the cube are identical. Nevertheless, we can define directions along the axes with a simple terminology. For example [100] is one unit cell distance a_0 along the x axis from the origin, with no displacement along the y and z axes.

The numbers in brackets in the upper panel of Fig. 2.12 indicate various directions in the unit cell. As discussed, they represent the distance (times a_0) that the origin must be translated in each direction to bring it to the desired point on the edge of the unit cell. It is also desirable to have some concept of crystal planes. The planes are a new idea, and we illustrate three such planes in the lower three panels of Fig. 2.12. Each of the three planes (shaded) have been given a designation, which are the *Miller indices* for those planes. How do we arrive at this description? The Miller indices of a plane are simply specified by giving the direction of a line perpendicular to the desired plane and pointing *out of the unit cell*. This choice of the direction of the line pointing into the cube is completely arbitrary, but it enables us to have different indices for the front and back faces of the cube.

Having defined the crystalline directions, there is an easy method of specifying the Miller indices. For this, we chose the plane of interest and determine its intersections with the three axes. In the lower left panel of Fig. 2.12, the plane intersects the x axis at 1 (which means one unit of a_0 along the x axis). It does not intersect the y or z axes, but common geometric interpretation puts these intersections at ∞ (that is, the plane is parallel to the plane formed by the y and z axes). Thus, the intersections of the plane are at the points 1, ∞, ∞. We now take the reciprocals of these intersections, or 1, 0, 0, and reduce them to the lowest common set of integers, which in this case is (100). These are the Miller indices of this plane and the commas are usually omitted. Consider an example in which these intersections are at $x = 2$, $y = 2$, and $z = 1$. The reciprocals are then 0.5, 0.5, and 1.0 along the three axes. These are then rationalized to the lowest set of integers, or 1, 1, 2. This would be the (112) plane.

It is important to point out that this definition suggested to use the lowest common set of integers, but this is quite arbitrary. It is as common to just use the reciprocals themselves if they represent a convenient set of integers. So, the last example could as easily have been referred to as the (224) plane, but this is not a usual reference *for this plane*. Consider, however, the plane with intercepts at 0.5, ∞, ∞. This is shown in Fig. 2.13. Taking the reciprocals yields the integers 2, 0, 0, and this is usually described as the (200) plane. In the BCC and FCC lattices, the (200) plane is quite important, as it corresponds to the atom in the body, or the atoms lying in the faces of adjacent sides of the cube. We will see this later when we look at the lattice image of the crystal. Another example is the back surface of the plane. Here, we normally shift the origin to a front corner, so that the intercepts of the back plane are at $x = -1$, $y = z \rightarrow \infty$, so that the reciprocals are -1, 0, 0. This is called the $(\bar{1}00)$ plane — the bar over the integer indicates the negative value of this integer.

The central lower panel of Fig. 2.12 shows a plane with intersections at $x = y = 1$ and $z = \infty$. The reciprocals of these are the points 1, 1, 0, so this is the (110) plane. Finally, the lower right panel shows

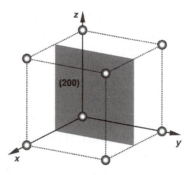

Figure 2.13. The (200) plane in a cubic unit cell. This plane intercepts the x axis half way along the length of the unit cell.

a plane with intersections of $x = y = z = 1$, for which the inverted numbers remain 1, 1, 1. This is the (111) plane. An important consequence of this is that the [111] direction in the crystal is the *normal* direction to the (111) plane. Similarly, the [100] and [110] directions are normal to the (100) and (110) planes, respectively. Hence, the Miller indices define a plane through its normal vector, where the latter is defined within the basic coordinate system of the lattice.

As a final example, consider the plane drawn in Fig. 2.14. Here, the intercepts are located at 1/2, 1/3, and 1/4 along the x, y, and z axis, respectively. The reciprocals of these intercepts are 2, 3, and 4.

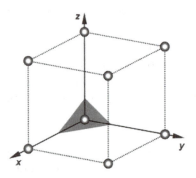

Figure 2.14. The (234) plane in a cubic unit cell. The intercepts of this plane occur at 1/2, 1/3, and 1/4 of the way along the unit cell in the x, y, and z directions, respectively.

The lowest common produce for which all three numbers are factors is 12, but it is normally the practice that we just use these integers to describe the plane as the (234) plane. This is not a very common plane, and one normally restricts themselves to the planes with higher symmetry, a point to which we return shortly.

To summarize, the Miller indices are commonly defined by taking the intercepts of the plane with the three axes (in a cubic crystal this is the three orthogonal directions). These are then inverted and converted to integers in the most convenient manner. Normally (but not always), this is the lowest set of integers. In general, this will result in the description of a plane as (hkl), where h, k, and l are integers.

The Miller indices have been denoted here using parentheses, whereas the crystalline directions have been denoted by brackets. There is no standard notation for these purposes, but we have attempted to utilize the most commonly used notation. Because of symmetry, certain sets of planes are often equivalent. These equivalent sets may be indicated by the use of braces, or curly brackets ({ }). This is true of the faces of the cube, and can be indicated as the set

$$\{100\} = (100), (010), (001), (\bar{1}00), (0\bar{1}0), (00\bar{1}). \qquad (2.4)$$

The (111) plane is just the plane which is perpendicular to the body diagonal. This plane passes through three nonadjacent corners of the cube. In zinc-blende materials, the (111) plane and the $(\bar{1}\bar{1}\bar{1})$ plane are *not* equivalent. Consider the semiconductor GaAs. In one case, we will see later that the plane normal points toward a Ga atom, while in the other it points toward an As atom. Since these two atoms are not equivalent, the two planes are not equivalent. Consider Fig. 2.15, where we have drawn a (110) plane of the zinc-blende lattice and indicated the atomic positions of the atoms in the unit cell. The significance of this plane is that all three primary directions [100], [110], and [111] lie in this plane. The unit cell is rectangular with length a_0 along [100] and $\sqrt{2}\ a_0$ along [110], which form the two perpendicular axes for this structure. The [111] direction points "diagonally" across the rectangle at an angle of approximately $35.26°$ with respect to the [110] direction. Along this [111] direction, one has, for example, a Ga atom at the lower left corner, then an As atom one quarter of the way along the "diagonal." These two atoms

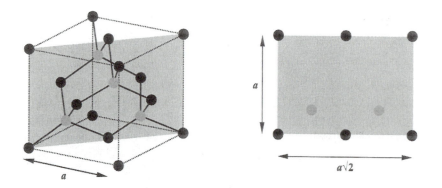

Figure 2.15. The (110) plane of the zinc-blende lattice. The left figure shows this plane within the context of the unit cell. The right figure shows the arrangement of different atoms within this plane.

actually form the *basis* for the lattice site at the lower left corner. Following this, there are no other atoms until reaching the Ga atom at the upper right-hand corner. If one takes a position midway between the two lower atoms of the basis at this lower corner, then looking along the [111] direction, we see an As atom. However, if we look along the [$\bar{1}\bar{1}\bar{1}$] direction, we see the Ga atom. The fact that the two atoms of the basis are different explains why these two directions are not equivalent. And, of course, the equivalent planes are not equivalent. If one says that the (111) plane at a surface is As atoms, called As terminated, then the ($\bar{1}\bar{1}\bar{1}$) plane would be Ga terminated. In Si, this becomes irrelevant, but it is important in the zinc-blende compounds.

Now, let us return to the lattice image, which we originally presented in Fig. 1.7. We show this again in Fig. 2.16(b), but now we have drawn some axes on the picture. The picture is made by imaging a thinned (perhaps 5- to 10-nm-thick) section of GaAs, with the (110) plane as the plane of the image. Using a high-voltage transmission electron microscope, one can image the actual lattice being studied. GaAs is a zinc-blende semiconductor with the FCC lattice and a basis of one Ga atom and one As atom per lattice site [as shown in Fig. 2.11, which is repeated as panel (a) of this figure]. The zinc-blende lattice differs from the diamond, in that the two atoms of the

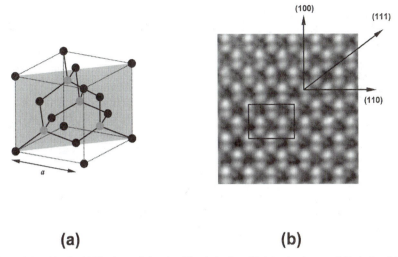

Figure 2.16. (a) The (110) plane of the zinc-blende lattice. (b) A lattice image of GaAs in which the different crystal directions are identified. The open box denotes the unit cell of GaAs in the (110) plane. *Picture provided courtesy of A. Trampert and O. Brandt, Paul-Drude Institute, Berlin.*

basis are the same for the latter structure (Si for example). The rectangular box in Fig. 2.16(b) represents the basic cell of the (110) plane. The vertical dimension is a_0, while the horizontal dimension is $\sqrt{2}\,a_0$. The six atoms that are connected by the black box are the six atoms connected by the rectangle in panel (a). These are four corner, and two face atoms (the top and bottom faces). These six atoms are, for example, all Ga atoms. There are two bright spots (as bright at the six Ga atoms), which seem to be centered in the rectangle. These are the other four face atoms, which are all Ga atoms. They overlay each other, with two being one layer up and the other two being one layer down. If the image is the (110) plane, the layer downward is the (220) plane. These two bright spots form triangles with the other six atoms, and in the center of each triangle is a less bright (gray) spot. These are the As atoms. The lower three (along one edge of the rectangle) Ga atoms, plus the adjacent two gray spot As atoms, form a zigzag chain, which can be seen in panel (a) along the bottom of the rectangle. The various crystal directions that lay in the (110) plane are also shown in Fig. 2.16(b) for reference.

2.4. X-RAY DIFFRACTION

In the previous sections, we discussed the nature of the crystalline structure of solids. How do we study this structure? In fact, there are many experimental methods of studying the crystalline properties of solids. Two of the most common are x-ray diffraction and electron diffraction. The two methods differ only in the probe signal used, x rays for the former and electron beams for the latter, and the equations of one can be rapidly and easily extrapolated to the other. However, to understand the similarity of the two, we need first to investigate the wave properties of the electrons, and this will be done in the next chapter. For this reason, we shall limit our discussion to x-ray diffraction, since it is readily accepted that the x rays are very high frequency electromagnetic radiation. Before beginning a discussion of crystal studies, however, let us set forth certain concepts clearly, so that we all understand what we mean by *diffraction.*

Suppose that we are considering the interference of a set of signal sources, or even a single signal source that must pass around a small obstacle. By small, we mean one whose size is smaller than, or comparable to, the wavelength of the radiation. *Diffraction* is the interference of a set of beams, which arise as scattered beams. This can either be the interference of the set of waves or of waves generated by passing on either side of the small obstacle (scattered by the obstacle). These beams can interact and produce either constructive or destructive interference. This process of interference is not limited to x-ray diffraction, and it is quite widespread. In the case of the crystalline solid, the scattering arises from the atoms of the solid itself.

If the waves in question originate by scattering from the atoms, then the phase differences relevant for diffraction arise from time delays due to the longer path lengths from each of the individual atoms. (In fact, the scattering of the x rays comes from interactions with the individual atoms, but we can consider the case of plane scattering.) Consider Fig. 2.17, where the incident radiation is reflected from the atoms located in different planes of the crystal. We have shown the case appropriate to either the BCC or the FCC lattice, in which the second and fourth planes are shifted with respect to the first or third planes. The lattice constant a_0 refers to the distance

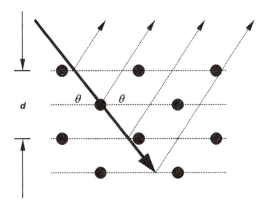

Figure 2.17. When an electromagnetic wave propagates through a crystal a small fraction of the incident intensity is reflected at each different crystal plane. The interference of these reflected waves is responsible for the phenomena of *Bragg diffraction.*

between either the even-numbered planes or the odd-numbered planes. Waves, which are reflected from these planes, interfere and lead to the crystal diffraction pattern that is observed. Then, one finds bright and dark spots in an image (which is typically recorded on photographic film) of the diffracted x rays. W. L. Bragg[1] found that he could explain the position of the diffracted beams by this simple method of diffraction from the solid. Consider that the x rays are reflected from adjacent planes of the crystal, as shown in Fig. 2.17. If the *equivalent* crystal planes are separated a distance a_0, then the difference of path length of the two beams is

$$\Delta L = 2a_0 \sin \theta. \qquad (2.5)$$

This gives a phase shift between the two beams of

$$\Delta \varphi = 2\pi \cdot \left(\frac{2a_0 \sin \theta}{\lambda} \right) = \left(\frac{4\pi a_0}{\lambda} \right) \sin \theta. \qquad (2.6)$$

Maxima of the diffracted beam should occur when the parallel rays are all in phase, or when

$$\left(\frac{4\pi a_0}{\lambda}\right)\sin\theta = 2n\pi, \qquad (2.7)$$

or

$$2a_0 \sin\theta = n\lambda. \qquad (2.8)$$

This is *Bragg's law*. Of course, to observe this, the source wavelengths must be such that $\lambda \ll 2a_0$. Normal lattice constants are on the order of a few tenths of a nanometer, and for this reason, very short wavelengths are required; hence x rays are used (see Table 2.1). For a lattice spacing of 0.5 nm, and a wavelength of 0.1 nm, the first Bragg angle is about 5°.

In Table 2.1, the appropriate x-ray wavelengths are shown for emission from the K shell of the atom (one of the inner shells). The K shell is the inner most atomic shell, as discussed in the next chapter. There are actually two closely spaced lines, which are called the α_1 and α_2 lines. The value given is the unresolved value that is commonly accepted. The x rays are produced by exciting the crystal with a high-energy beam of electrons. These drive electrons out of the shell, and when they relax back to their normal bound state, x rays are emitted. Hence, the frequency depends on the material, through its atomic energy level structure, and on which shell is excited.

Let us consider an example of how Bragg diffraction works. From Fig. 2.17, we note that the angle of incidence of the beam is the same

TABLE 2.1. X-ray Wavelengths

Material	Kα wavelength (nm)
Ag	0.05684
Mo	0.071073
Cu	0.154184
Ni	0.165919
Co	0.179026
Fe	0.193735
Cr	0.2291

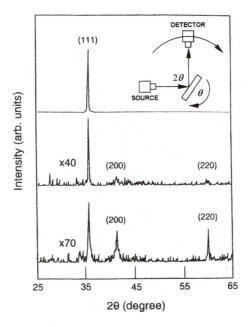

Figure 2.18. X-ray diffraction patterns of SiC films prepared at different substrate temperatures (top, 800°C; middle, 900°C; bottom, 1000°C). The peak marked (111) is associated with the periodicity of the (111) plane in silicon. *Y. Sun et al., J. Appl. Phys.* **82**, *2334 (1997).Figure reproduced with the permission of the American Institute of Physics.*

as the exit angle of the reflected radiation. Normally, it is quite difficult to move the x-ray source, so the sample is rotated by the angle θ and the detector is rotated through an angle 2θ. This is depicted in the inset of Fig. 2.18. The detector signal is then plotted as a function of this 2θ, and a typical trace is shown in the main panel of Fig. 2.18. Here, the signal is for cubic SiC grown on a Si substrate. The important point is the spacing of equivalent planes. If we describe a plane by the Miller indices (hkl), it is easily shown that we replace a_0 with d given by

$$d = \frac{a_0}{\sqrt{h^2 + k^2 + l^2}}.$$
(2.9)

We note that there is a major peak at $2\theta \sim 35.8°$, which gives a value of the angle of $\theta = 17.9°$. This peak is marked as the (111) plane for the underlying Si substrate. Using the wavelength of the x-ray source as that of the Fe Kα line (see Table 2.1), or 0.19 nm, we arrive at a value of $d = 0.315$ nm for this peak. However, this is along the body diagonal, which is actually $a_0/\sqrt{3}$. Making this correction, $a_0 = 0.546$ nm. The proper value for Si is 0.542, so that the error is within 1% (which is certainly within our reading of the figure itself). The second peak arises from the SiC, and is identified as the (200) peak, which involves one-half of the cubic a_0 value for SiC. This angle is $2\theta \sim 41.2°$, which gives $\theta = 20.6°$, for which we find the cubic $a_{0,SiC} = 0.55$ nm. This is slightly larger than that of Si, and within the range expected from the growth process.

In actual fact, Bragg diffraction is more properly a reflection since the incident angle is equal to the exit angle. The waves are reflected from each plane. Yet, the interference that arises from the multiple planes makes this a diffraction phenomenon due to the constructive and destructive manner in which this interference can occur. In general, diffraction can be more complicated. Consider Fig. 2.19, where we sketch a line of atoms (or planes). The incoming angle depends only on the angle of the source relative to the plane, whereas the outgoing angle depends also on the properties of the diffracting atoms (or planes). Figure 2.19 shows two parallel ray paths for the light. These are defined by being normal to the plane of the wave. The phase difference between sources for the outgoing wave is

$$\alpha + \left(\frac{2a_0\pi}{\lambda}\right)\sin\theta_{out}, \qquad (2.10)$$

where α is the phase shift of the radiation at each atom. In general, this phase shift is given by the exciting, or input beam, so that

$$\alpha = -\left(\frac{2a_0\pi}{\lambda}\right)\sin\theta_{in}. \qquad (2.11)$$

The minus sign appears on θ_{in} due to the opposite direction of the arriving rays. Then the phase shift in the two out-going waves is

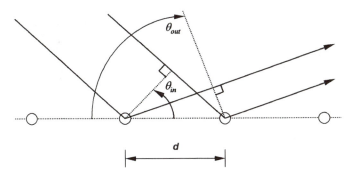

Figure 2.19. Light diffraction.

given by

$$\Delta\varphi = \frac{2a_0\pi}{\lambda}(\sin\theta_{out} - \sin\theta_{in}). \tag{2.12}$$

Diffraction differs from reflection (the Bragg case), in that for reflection $\theta_{in} = -\theta_{out}$ (the minus sign arises from the difference in the definitions between Figs. 2.17), while in diffraction the magnitudes of these two angles can be different. This has led to a more general description of x-ray diffraction, which is due to von Laue.[2,3] The maxima of the diffracted beam occurs when the phase shift is $2n\pi$, where n is an integer, just as in (2.7). Then

$$d(\sin\theta_{in} - \sin\theta_{out}) = n\lambda, \tag{2.13}$$

which is slightly different from (2.8). The important difference between this result and the Bragg result is that there are a series of angles for which the diffraction pattern will occur. Especially in three dimensions, the pattern of light spots will show the symmetry of the crystalline arrangement. Note, however, that the angles defined in Fig. 2.18 are not the same as those of the Bragg set — they are defined from the plane instead of from the normal to the plane.

Scattering by x rays using the von Laue equation (2.13) is typically done by irradiating with an x-ray beam and recording the diffracted output beams on a photographic plate, as shown in

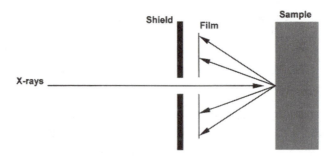

Figure 2.20. Experimental arrangement for the von Laue back reflection x-ray diffraction.

Fig. 2.20. In this approach, we measure all of the diffracted beams at the same time. However, we need to rotate the axis of reference for the angles, and to do this we replace $\theta \rightarrow \pi/2 - \theta$. Using this change in (2.13) now gives the Laue equation as

$$a_0(\cos \theta_{in} - \cos \theta_{out}) = n\lambda, \qquad (2.14)$$

except that now the angle is measured from the normal to the plane, just as in the Bragg case. Now, consider a cubic crystal with the x rays incident parallel to the x axis of the crystal [normal to a (100) plane], and we take the x-ray wavelength to be 0.154 nm, corresponding to the Cu Kα line. With respect to the (100) plane, the angle of incidence is $\theta_{in} = 0$. Now, unfortunately due a set of circumstances in the diamond structure, neither the (100) nor the (200) plane will cause a diffraction pattern (there is a phase cancellation that is beyond the present treatment to explain — it is just an experimental fact). Now, consider the (220) plane, again with the (x, y) arrangement. While the beam is still normal to the face of the crystal, it makes an angle of $\theta_{in} = 45°$ with respect to the (220) plane. Moreover, the spacing of the planes is reduced to $d/\sqrt{8}$. As a result, the beam now comes out at an angle of $\theta_{out} = 95.5°$ relative to the rotated axis. After rotating back to the reference axis, the angle is 50.5°. There are four of these beams, and they form a fourfold symmetry relating to the symmetry of the crystal. Quite simply, the use of the back-scattered x-ray spots (on the photographic plate) enables a determination of the crystal structure of the material under study. By measuring the angles, one can ascertain the symmetry planes that are allowed in the crystal, and the lattice constant itself. The number of beams and their orientation tells us about the symmetry of the crystal so that it is possible from an extensive investigation to determine quite a lot about the crystal.

The significance and importance of x-ray scattering was not lost on these early scientists. Max von Laue received the Nobel prize in 1914 for his work on x-ray scattering, and Bragg received the prize in 1915 for his work on crystallography. Considering that x rays themselves had only been discovered a few years earlier, this was a remarkable advancement in the characterization of materials.

REFERENCES

1. W. L. Bragg, Proc. Cambridge Phil. Soc. **17**, 43 (1913).
2. M. von Laue, Ann. Physik **41**, 989 (1913).
3. W. Friedrich, P. Knipping, and M. von Laue, Ann. Physik **41**, 971 (1913).

PROBLEMS

1. Sketch the (211), ($\bar{2}$11), and (311) planes for a cubic crystal.
2. What are the frequencies for the x rays shown in Table 2.1?
3. Show that the van Laue equations in one dimension reduce to Bragg's law.
4. Develop the equivalent set of von Laue equations for three dimensions. In particular, establish the validity of (2.9).
5. The lattice constant of germanium is 0.566 nm. For radiation of 0.025 nm, find the first Bragg angle for reflection.
6. Consider the von Laue equations in three dimensions found in Prob. 4. If a crystal is oriented so the x ray is incident parallel to one of the cube edges (for a cubic lattice), where are the {111} spots?

CHAPTER 3

The Wave Mechanics of Electrons

Just over 100 years ago, the understanding of classical physics was nearly complete—or so it was thought. Particle motion was fully described by Newton's equations of motion, and light (an electromagnetic wave) was described fully by Maxwell's equations. Particles were not waves, and waves were not particles, so these two sets of equations were not coupled. Only waves could show interference phenomena, such as those used in Chapter 2 to characterize crystals. Yet, there were disturbing experimental situations. For example, it was known that accelerated particles as well as particles moving at constant velocity in circular orbits could generate electromagnetic waves (which would lead to the collapse of the electron energy in classical atoms). This was not a cause for concern about the distinct nature of particles and waves, however. Other unexplained phenomena would come to change this view.

The common understanding assured us that things like electrons, ball bearings, tennis balls, and even the planets were decidedly particles—even great assemblages of individual particles were particle-like in nature. On the other hand, light, heat, x-ray radiation, and water had decidedly wavelike behavior. The fact that water was a fluid composed of a great many individual particles did not seem to count for much — water displayed waves, as any fool could see by merely taking a trip to the beach. However, the wave equation for water was not one described by Maxwell's equations; but this was not thought to be a cause for concern. Waves clearly had certain properties, such as wave height, velocity of propagation, and wavelength, the latter being the distance between two successive troughs or peaks. Light waves were taken to have the same properties — amplitude (instead of height), velocity of propagation, and wavelength. Indeed, some of these quantities were related — the frequency f, the velocity of propagation c, and the wavelength λ were related through

$$f = \frac{c}{\lambda}. \tag{3.1}$$

For example, red light has a wavelength near 0.8 μm, which, according to (3.1), corresponds to a frequency $f = 3.75 \times 10^{14}$ Hz. While water possesses a velocity that depends on the fluid density and other factors, electromagnetic waves, such as light and x-rays, clearly propagated at the speed of light,* $c = 2.997923 \times 10^{10}$ cm/s.

The beginning of the end of this comfortable situation arose with the resolution of the blackbody radiation problem. Every physical body radiates electromagnetic waves, either in the visible spectrum or in the infrared spectrum as heat. It was known that the intensity of this radiation was frequency dependent. The problem was that there were two incompatible descriptions of this radiation. Lord Rayleigh found that at low frequencies, the energy density of the radiation varied as the square of the frequency through**

$$\rho_E(f) \sim f^2 T \, df. \tag{3.2}$$

If this were integrated with a Maxwellian weighting function, it would give the well-known Stefan–Boltzmann law in which the

---

*Albert A. Michelson[1] began his measurements of the speed of light while an officer in the U.S. Navy. Although born in Germany, he attended the Naval Academy and returned there to teach until 1880. He was the first American to receive the Nobel prize, when he was awarded the 1907 Physics prize for these measurements, first done in 1878 while at Annapolis.

**This is discussed nicely in Longair.[2]

radiation varies as T^4, where T is the temperature of the emitting body. However, this relation did not hold at high frequencies, and it was suggested by Wien that the energy density of the radiation varied as

$$\rho_E \sim f^3 e^{-Af/T}\, df. \tag{3.3}$$

Attempts to reconcile these two formulas failed until around 1900. Then, Planck made the daring assumption that the energy could be emitted only in discrete quantities, which were termed *quanta*.[3] Instead of radiation being a continuous phenomena, it was now to be considered almost particlelike, with little packets of energy being spewed forth from the blackbody's surface. These particles are called *photons*. Even Planck found this concept to be unsettling, and he tried not to emphasize the radical nature of this new assumption. Yet, he initiated the quantum world, with the famous equation

$$E = hf, \tag{3.4}$$

where h was a new, fundamental constant with the value $h = 6.62618 \times 10^{-34}$ J-s. (The more common form in which it is seen moves a factor of 2π from h to f, creating the radian frequency ω, so that we have the reduced Planck constant $\hbar = h/2\pi = 1.05459 \times 10^{-34}$ J-s.) Planck's blackbody radiation law gave us the final formula:

$$\rho_E(f) \sim \frac{f^3}{e^{hf/k_BT} - 1}. \tag{3.5}$$

For large f, this gives the Wien result, while for small f, the denominator can be expanded and yields the Lord Rayleigh result. While this new idea of quanta was difficult to accept, it was only a few years until it was clearly validated by Einstein's theory of the photoelectric effect. In Fig. 3.1, we plot (3.5) for three different temperatures of the emitting surface. This emission is that which is seen from all thermal light sources, whether they be the sun or a light bulb. We encounter this temperature even when we talk about the color spectrum of a television display.

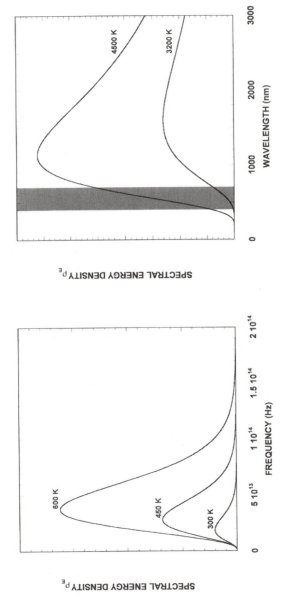

Figure 3.1. Spectral energy density of blackbody radiation. Shown on the left is the spectral density as a function of frequency, whereas shown on the right is the spectral density plotted as a function of wavelength. In the latter figure, the gray region represents the visible portion of the electromagnetic spectrum, between 400 and 700 nm.

3.1. THE PHOTOELECTRIC EFFECT

The next problem came from the observation that a solid could emit particles as well as photons, especially when photons entered the solid. The photoelectric effect arises when a solid body is illuminated with light. If the light has a sufficiently high frequency, then electrons are emitted from the body. This was the first peculiar fact about the photoelectric effect — it isn't how intense the light beam is, but the frequency of the light wave that determines whether or not electrons are emitted. Moreover, if one increases the intensity of the light wave, once electrons are being emitted, the energy of the emitted electrons *does not increase*! Rather, more electrons are emitted. Thus, the photoelectric effect was quite strange, and no one had explained why these observations were expected. Then, in 1905, the young Albert Einstein published a paper explaining the photoelectric effect in terms of the quanta introduced by Planck (it was this for which Einstein received the Nobel prize in 1921, not for his more famous relativity theory).[4]

If the light waves are interpreted as a beam of quanta, each with an energy given by (3.4), then the interpretation of the photoelectric effect is straightforward. The energy of these quanta, known as photons, is given by (3.4) and is independent of the intensity of the light radiation. Instead, the intensity just increases the *number* of photons that arrive each instant of time. The key connection is the critical frequency f_{cr} required for an electron to be emitted. The energy corresponding to this critical frequency is known as the *work function*

$$E_W = hf_{cr} = \hbar\omega_{cr}. \tag{3.6}$$

It was known that the kinetic energy of the emitted electrons was given by the light frequency, and the work function, as

$$E = hf - E_W, \tag{3.7}$$

and this can be interpreted as a connection between the emitted electron and a barrier. That is, the work function can be seen to be a barrier between the electrons in the solid and the free space where the photons emanate. Then the energy of the incoming photon is split

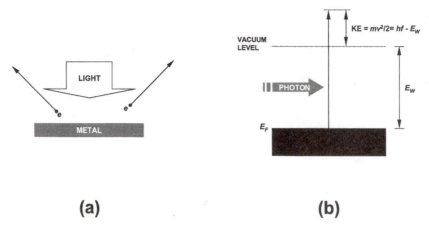

Figure 3.2. (a) In the photoelectric effect, light is shone on a metal and electrons are emitted from its surface. (b) Relevant energy scales for a discussion of the photoelectric effect. (*See the main text for a definition of the different symbols used.*)

between the kinetic energy of the emitted electron and its potential energy is required to raise it to the height of the barrier. Only the kinetic energy of the emitted electrons was being measured. This is shown in Fig. 3.2. The incoming photon excites an electron from the filled states below the Fermi level (the highest filled state in a metal — we treat the Fermi level in more detail in Chapter 5). A portion of the photon's energy goes in raising the electron to the top of the barrier, and the remainder can then appear as kinetic energy (we deal with this in Secs. 3.4–3.5), and the particle's velocity can be found from

$$\tfrac{1}{2}mv^2 = hf - E_W. \tag{3.8}$$

The remaining observation — that the number of emitted electrons increased with the intensity of the light beam — is explained by (3.8) as well. Since the emission of the electron is intimately connected to the arrival of a photon, increasing the number of photons simply increases the number of electrons proportionally. Thus, the current emitted is directly related to the intensity of the light wave

$$n_{\text{em}} = \frac{CI}{hf}, \tag{3.9}$$

where I is the intensity of the light wave and C is a proportionality constant. The intensity may be decomposed into the number of photons times the power carried by each photon. The latter is just the energy hf times the velocity of propagation c. Thus, we may write $I = n_{ph}hfc$ W/cm^2. The number of emitted electrons is related to the number of photons through

$$n_{em} = Ccn_{ph}. \qquad (3.10)$$

The speed of light is often included with C into a new constant, which is termed the quantum efficiency.

The explanation of the photoelectric effect firmly established that waves could actually be interpreted as particles, in this case the light wave is composed of photons. This fact would be one of the driving forces for the later development of the new quantum theory. Another driving force would be the Bohr model of the atom.[5,6] In this case, Bohr explained the observation of specific light emission frequencies from atoms by a model in which the electrons existed in orbits with specific energies. It was the differences in these energies that corresponded to the observed light frequencies. In creating the model, he had to abandon the classical notion that accelerated particles would radiate their energy (the particles moving in curved orbits would be assumed classically to give off radiation and thus slow down). The fact that the particles could have only certain energy values meant that this energy was quantized, just as the photons. This faced scientists with an important question: If light waves can be treated as particles, does the converse hold? That is, can particles such as electrons be treated as waves?

3.2. ELECTRONS AS WAVES

It is well accepted that light waves can be diffracted and show interference fringes. What about electrons or other particles? In fact, it was discovered quite soon after Einstein's description of the photoelectric effect that electrons would show the same diffraction and interference phenomena. The classic optical wave-interference

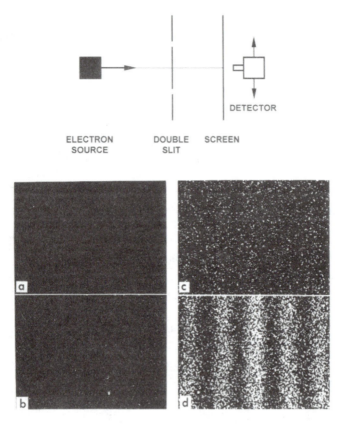

Figure 3.3. Buildup of the double-slit interference pattern in an experiment in which electrons are used as the wave source. Shown top is a schematic diagram illustrating the main components of the experimental apparatus. Shown below are photographic images obtained after different numbers of electrons have passed through the apparatus: (a) 8 electrons, (b) 270 electrons, (c) 1600 electrons, and (d) 70,000 electrons. *A. Tonomura, Advances in Physics* **41**, *59 (1992). Figure reproduced with the permission of Academic Press.*

effect is caused by the passage of light through two slits. If we devise a system in which light emanates from a well-defined source, and illuminates a plane in which two slits are cut, then on another plane behind this first one, there will appear an interference pattern. Such an interference pattern is a sequence of light and dark spaces corresponding to constructive and destructive interference of the light passing through the two slits. If either of the slits is covered, the interference pattern goes away. Such an experiment is shown in Fig.

3.3 for electrons. In addition, the pattern on the second screen is shown as a large number of electrons passes through the two slits. For just a few electrons, no pattern is visible, which is also the case for only a few photons. After a great many electrons have passed through the two slits, however, the interference phenomenon is well described in the pattern on the second screen. This is a most unusual result!

From these experiments, we are led to conclude that all things, whether particles or photons or automobiles, are *both* particle- and wavelike in nature. Which property we measure depends on the type of measurement being performed. The two-slit experiment is a wavelike measurement, as it depends on the interference of two waves emanating from the slits themselves. On the other hand, the photo-electric effect is a particlelike experiment, as it depends on the energy and momentum of particles, whether photons or electrons. In essence, any object has a dual explanation — either as a wave or as a particle. We summarize this in talking about wave–particle *duality*. A physical body has both wavelike and particlelike properties, and we can describe it by either of these properties. When a measurement is made, however, the measurement will select which of the properties is important to that specific measurement. To be sure, this view is completely different from the simple ideas of classical mechanics. To bring this to a better understanding, however, some holes must be filled in.

Consider, for example, the wavelike description. For a photon described by its wave, there is a well-defined wavelength, given by (3.1) in terms of the frequency, and hence of the energy. How are we to arrive at the wavelength for the wave properties of an electron? The answer was given by Louis de Broglie. De Broglie first raised the wavelike interpretation by suggesting that the quantization of an electron around an atom came from the need to have a fixed number of wavelengths in each orbit.[7] In this paper, de Broglie assigned to an electron a fictitious wave which had a frequency f and wavelength λ. These quantities were presumed to be related by the Planck condition (3.4) and

$$p = mv = \frac{h}{\lambda}. \tag{3.11}$$

The quantity p is the momentum of the electron. Since $E = p^2/2m$ for a freely moving particle, the energy is now related to the momentum by

$$E = \frac{h^2}{2m\lambda^2},$$ (3.12)

which is clearly different from that expected from (3.1) and (3.4) ($E = hf = hc/\lambda$ for the photon). The difference lies in the fact that *particles do not travel at the speed of light*, and their velocity varies with the energy. The wavelength given by (3.11) is known as the *de Broglie wavelength*, and it applies to the wave description of particles, which do not travel at the speed of light.

In the preceding discussion, the wave frequency and wavelength are related to the energy and momentum of the corresponding particle by (3.4) and (3.11), respectively. Moreover, the treatment of a particle as a wave — and of a wave as a particle — is a general property. Which treatment is to be used depends on the constraints of the problem. Two major problems are left, however. The first concerns the nature and properties of the waves. The second concerns the manner in which the wave treatment is applied to physical problems. These are called problems because the wave treatment is foreign to our present understanding of classical mechanics. We now consider the first of these problems, and return to the second later. For an oscillatory wave, the velocity at which a plane of constant phase angle moves is called the *phase velocity*, and may be defined as

$$v_{\text{phase}} = \frac{\omega}{k},$$ (3.13)

where $k = 2\pi/\lambda$ is the wave number for the wave. However, the velocity at which the wave *energy* travels may not be the same as the phase velocity. If a localized wave — or a wave whose amplitude is localized to a narrow region of space — is chosen, then the amplitude of the localized wave may be related to the position of the particle. The velocity of the particle is related to the velocity at which the localized wave amplitude moves. Such a localized wave is called a *wave packet*, and the velocity of this packet is the *group velocity*. In general, the group velocity of the wave packet is not the same as the

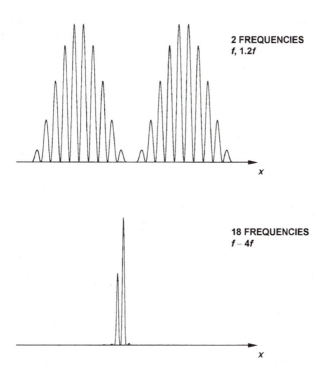

Figure 3.4. Waveforms obtained by adding sinusoidal waves with a number of different frequencies together. Note that in this case, the *squared* amplitude of the wave is plotted. The formation of a wave packet for 18 frequencies is clearly seen.

phase velocity. In Fig. 3.4, we plot the sum of waves that compose a wave packet. We can see that it is the additive (constructive) interference of these waves that produces the localized packet. Media in which the two velocities are different for simple waves — for example, a single sinusoid — are called *dispersive media*. The general property that ascribes different values to the phase and group velocities is known as dispersion.

To investigate the dispersion of a wave packet, let us consider a situation in which the particle — an electron, for example — moves in rectilinear motion in the x direction. The accompanying wave clearly oscillates in space as it travels with the particle. Such a wave can be expressed as

$$\psi(x, t) = A \sin \theta(x, t), \tag{3.14}$$

where A is some constant amplitude and θ is the phase relationship of the wave; θ varies in such a manner that the planes of constant phase move at a velocity v_{phase}. Thus, at some point x, a plane of constant phase moves such that

$$\frac{dx}{dt} = v_{\text{phase}}. \tag{3.15}$$

Using (3.13) in (3.15) gives us

$$\frac{dx}{dt} = \frac{\omega}{k}, \tag{3.16}$$

or

$$\theta = kx - \omega t. \tag{3.17}$$

If we differentiate (3.17) with respect to t, we arrive at (3.16). If x is held constant, the wave oscillates in time at a frequency $f = \omega/2\pi$. While if the time is held constant, the wave has successive nodes at distances separated by $\lambda/2$ (the wave is spatially periodic with period λ). That is, successive nodes appear at values of x given by

$$x_n = \frac{n\lambda}{2} + v_{\text{phase}}t. \tag{3.18}$$

Thus, the nodes travel at a speed given by (3.15).

In a more proper sense, though, a wave packet localized near a particular point in space must be used if this wave packet is to represent the particle. Such a packet, or group of waves, can be demonstrated by first adding two waves to obtain a single composite wave. Take, for example,

$$\begin{aligned}
\psi_1 &= \sin(kx - \omega t) \\
\psi_2 &= \sin[(k + \Delta k)x - (\omega + \Delta\omega)t].
\end{aligned} \tag{3.19}$$

That is, the second wave has a slightly different wave number (wavelength) and frequency. The composite wave is found by adding these two together as

$$\psi = \psi_1 + \psi_2$$

$$= 2\cos\left(\frac{\Delta k}{2}x - \frac{\Delta\omega}{2}t\right)\sin\left[\frac{2k + \Delta k}{2}x - \frac{2\omega + \Delta\omega}{2}t\right]. \quad (3.20)$$

The second term is rapidly oscillating with nearly the same propagation parameters as the two individual waves. However, the amplitude has a very slow envelope variation, as was shown in Fig. 3.4 for such waves. The envelope moves in space and time as

$$\frac{d}{dt}\left(\frac{\Delta k}{2}x - \frac{\Delta\omega}{2}t\right) = \frac{\Delta k}{2}\frac{dx}{dt} - \frac{\Delta\omega}{2}. \quad (3.21)$$

If we arbitrarily chose the points where this equation is zero, we find that these move with the new velocity

$$\left.\frac{dx}{dt}\right|_{\text{amplitude}} = \frac{\Delta\omega}{\Delta k} \rightarrow \frac{d\omega}{dk} \equiv v_{\text{group}}. \quad (3.22)$$

In this last expression, we have defined the *group velocity* as the velocity at which the packet amplitude moves.

Now, consider (3.12) for electrons. We can rewrite this as $E = hf = h^2/2m\lambda^2$, or $\omega = \hbar k^2/2m$. From this, we can find

$$v_{\text{phase}} = \frac{\hbar k}{2m}, \qquad v_{\text{group}} = \frac{\hbar k}{m}. \quad (3.23)$$

We will see that the second velocity is related to that of the particle, as the momentum is connected with the quantity $\hbar k$. This follows, as the classical definition of the velocity is given as

$$v = \frac{dE}{dp} = \frac{d(\hbar\omega)}{d(\hbar k)} = \frac{d\omega}{dk}. \quad (3.24)$$

Thus, the group velocity is that assigned to the particle represented by the wave, and this is closely connected to the momentum. As an example, let us consider an electron, whose mass is 9.1×10^{-31} kg, with a velocity of 10^7 cm/s. The momentum of this electron is $mv = 9.1 \times 10^{-26}$ kg-m/s, which corresponds to a de Broglie wavelength of approximately 7×10^{-9} m. At the same time, this electron has an energy of 4.55×10^{-21} J joules or 28.44 meV.

3.3. THE SCHRÖDINGER EQUATION

The physical interpretation of the wave function that was introduced in the previous section is simple. Bohr postulated that since the group of waves (or wave packet) is assumed to move with the particle, it must represent the probable position of the particle. More exactly, the square of the magnitude of the wave function is the probability density of finding the particle at a given position in space. That is, the probability of finding the electron (or particle) in a region of space $\Delta x \Delta y \Delta z$ centered about x, y, and z is given by

$$|\psi(x, y, z)|^2 \Delta x \Delta y \Delta z. \tag{3.25}$$

In Fig. 3.5, we plot a typical one-dimensional Gaussian wave packet in panel (a), and then in panel (b), we plot the probability of finding the particle at a position less than x. There is an important distinction between the *probability density* and the *probability* that is plotted in Fig. 3.5(b). In Fig. 3.5(a), the wave function that is plotted can be written as

$$\psi(x) = A \exp\left[-\frac{(x - x_0)^2}{2\sigma^2} \right], \tag{3.26}$$

where σ is the half-width of the Gaussian probability density. The *probability density* is just the square of the magnitude (which here is the square of the wave function as our Gaussian is a real function) and is given by

$$|\psi(x)|^2 = A^2 \exp\left[-\frac{(x - x_0)^2}{\sigma^2} \right]. \tag{3.27}$$

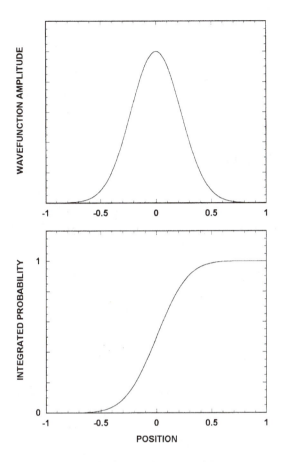

Figure 3.5. Top: wave function variation for an arbitrary particle. Bottom: *Integrated* probability of finding the particle in the range between $-\infty$ and x [see Eqs. (3.26) to (3.28)].

The function given by (3.27) is a function that is peaked around x_0, so that this latter quantity is the most likely value of x. We call this the *expectation* (or average) value of x and denote it as $\langle x \rangle = x_0$. On the other hand, the *probability* is the integral of (3.27), so that the probability of finding the particle at a position less than x is given by

$$P(x) = \int_{-\infty}^{x} |\psi(x')|^2 \, dx' = A^2 \int_{-\infty}^{x} \exp\left[-\frac{(x' - x_0)^2}{\sigma^2} \right] dx'. \quad (3.28)$$

It is this probability that is plotted in Fig. 3.5(b). The importance of this probability lies in taking the limit $x \to \infty$. Then, the probability that the particle is somewhere is unity. It must be at some point in space. By integrating over all space, we must have included the particle. This means that the limit of (3.28) is 1. This establishes a *normalization condition* for our wave function. We may restate this as

$$\int_{-\infty}^{\infty} |\psi(x)|^2 \, dx = 1. \tag{3.29}$$

In the case of (3.26) and (3.27), this latter result means that the constant A must have a very specific value. Using (3.29), we find that (3.26) can now be written as

$$\psi(x) = \frac{1}{\sqrt{\pi\sigma}} \exp\left[-\frac{(x - x_0)^2}{2\sigma^2} \right]. \tag{3.30}$$

Now, how does one arrive at this wave function?

In classical mechanics, simple equations of motion are used to solve for the particle position. In quantum mechanics, however, there is an equation of motion that yields the movement of the wave packet, and the solution to this equation is a more complicated process. Before we can even begin to solve the equation, we must first find the equation itself. *In fact, there is no way in which this equation can be derived.* The equation is known as the *Schrödinger equation*, and we can follow a rational argument that will give us the equation. But, we must realize that we know what the equation is supposed to be, and this argument just gives us the desired answer. We consider waves, such as those discussed in the previous section. These waves in general can be taken to have the same propagating nature as electromagnetic waves, so that we will assume a priori that the waves have a factor that depends on the wave number k, the position x, the frequency ω, and the time t as

$$\psi \sim e^{i(kx - \omega t)}. \tag{3.31}$$

We require the wave packet to satisfy the law of conservation of energy. The total energy of the particle is just

$$E = T + V \tag{3.32}$$

where T is the kinetic energy and V is the potential energy. The potential energy results from any applied force, so it will be just some function of position. On the other hand, the kinetic energy arises from the dynamics of the particle, so we can write this quantity as

$$T = \frac{p^2}{2m} = \frac{1}{2m}\left(\frac{h}{\lambda}\right)^2, \qquad (3.33)$$

where the de Broglie relation (3.11) has been used to replace the momentum p. This step is where we first introduce quantum behavior with the use of the de Broglie wavelength to describe the particle. It is at this point that we assert the particles are to be described as waves. We can now use the relationship between the wavelength and the wave number ($\lambda = 2\pi/k$) to arrive at

$$T = \frac{\hbar^2 k^2}{2m}. \qquad (3.34)$$

At this point, we finally use the assumed form of the wave (3.31) to replace the wave number with the derivatives of (3.31). That is, we make the assumption that the parameter k arises from (3.31) by taking a derivative with respect to x. As a result, the kinetic energy becomes a differential operator[†]

$$T \rightarrow -\frac{\hbar^2}{2m}\frac{\partial^2}{\partial x^2}. \qquad (3.35)$$

It is important to note that the k^2 is taken as $k \cdot k$, so that the differential operators (the first derivatives) are imposed sequentially to yield the second derivative and *not* as a product (which would have given the square of the first derivative). We now impose the wave properties a second time by using the Planck relation (3.4), but

---

[†]*Operators*, in general, represent an action taken on the wave function that produces a result. The potential simply multiplies the wave function, but the kinetic energy term involves operating on the wave function with the differential operators — taking the second derivative of the wave function.

with the 2π moved to the frequency, for the total energy

$$E = \hbar\omega \rightarrow i\hbar\frac{\partial}{\partial t}. \tag{3.36}$$

In the last expression, the transition from frequency to the differential operator for the time variation has again used the form (3.31).

The statement of the energy has now taken on differential operators, which require a function on which to operate. This function is, of course, the wave function that we express using (3.32) as

$$T\psi + V\psi = E\psi. \tag{3.37}$$

If we now insert the results (3.35) and (3.36), we arrive at the *Schrödinger equation* for the wave function

$$-\frac{\hbar^2}{2m}\frac{\partial^2\psi(x, t)}{\partial x^2} + V(x)\psi(x, t) = i\hbar\frac{\partial\psi(x, t)}{\partial t}. \tag{3.38}$$

As mentioned, this approach is not a derivation, but rather is a justification. The quantum behavior was introduced through assuming both the de Broglie wavelength and the Planck quantization condition. The wave nature itself was introduced by the assumed variation given in (3.31). The result (3.38) is a *diffusion* equation, in that it is second order in spatial derivatives and first order in time derivatives. (Such diffusion equations occur widely in classical physics, and we will encounter them again in later chapters. Here, it is important to note that the quantum waves satisfy a diffusion equation, while classical waves satisfy a wave equation — the equation is second order in the time derivatives.) Indeed, the combination of constants $\hbar/2m$ has the units of a diffusion constant (cm^2/s). However, the equation is complex, particularly because of the factor of $i(=\sqrt{-1})$ preceeding the time derivative.

The equation can be separated into its temporal and spatial parts, so that each part can be solved separately. To achieve this, assume

$$\psi(x, t) = \psi(x)\chi(t). \tag{3.39}$$

If we insert this form into (3.38), and divide by the wave function, the result is

$$\frac{1}{\psi(x)}\left[-\frac{\hbar^2}{2m}\frac{\partial^2\psi(x)}{\partial x^2} + V(x)\psi(x)\right] = \frac{i\hbar}{\chi(t)}\frac{\partial\chi(t)}{\partial t}. \qquad (3.40)$$

The term on the left-hand side of (3.40) is a function only of x, while the term on the right-hand side is a function only of t. The only way in which they can remain equal to one another as x and t vary is for each to be equal to a constant. The right-hand side was obtained from the total energy E, and this is a constant of the motion. As a result, we obtain the *time-independent* Schrödinger equation

$$-\frac{\hbar^2}{2m}\frac{\partial^2\psi(x)}{\partial x^2} + V(x)\psi(x) = E\psi(x). \qquad (3.41)$$

This gives the spatial variation of the wave function when there is no time variation, or when this time variation is independent of the spatial variation [which was assumed in using (3.39)]. At the same time, the time variation is found to arise from

$$i\hbar\frac{\partial\chi(t)}{\partial t} = E\chi(t), \qquad (3.42)$$

or

$$\chi(t) = \chi(0)e^{-iEt/\hbar} = \chi(0)e^{-i\omega t}. \qquad (3.43)$$

In the last expression, the Planck relation has once again been utilized to recover the actual initial assumption (3.31) about the wave.

In introducing the differential operator in (3.35), a crucial change in the nature of the momentum was made. In classical mechanics, both position and momentum are simple functional properties of the particle. Here, however, the position seems to be a simple property of the wave function representation of the particle, rather than the particle itself. The momentum has clearly changed. Instead of a simple functional property, the momentum is now a differential operator

$$p = \hbar k \rightarrow -i\hbar \frac{\partial}{\partial x}.$$ (3.44)

The first equality is just another statement of the de Broglie wavelength and the second result is arrived at by using (3.31).

The solutions that we obtain from the quantum-mechanical Schrödinger equation differ from those obtained with classical mechanics. In the latter, the solution gives the precise position and velocity of particles. With this quantum-mechanical equation, we find only a probability wave function, and this tells us only where a particle is *likely* to be. There exists an *uncertainty* as to the precise position and velocity of the particle. An inescapable result is that, in the wave treatment, one cannot determine simultaneously the exact position and the exact momentum of the electron. But this is what one might infer if one tried to make such a measurement. Measuring the velocity, or momentum, implies measuring a distance traveled in a given time interval. Because of the nonzero distance involved, one cannot specify the exact position that corresponds to the measurement. It is unfortunate that this contradicts what is generally taught in classical mechanics. Quantum mechanics states that this inability to measure both the momentum and the position exactly is not just a result of the method of measurement — as classical mechanics would infer — but is a fundamental theoretical limit, which cannot be overcome by any measuring system. In some sense, this is a result of the introduction of the operators. Let us consider the relationship

$$-i\hbar \frac{\partial}{\partial x}(x\psi(x)) = -i\hbar\psi(x) - i\hbar x \frac{\partial \psi(x)}{\partial x}.$$ (3.45)

This can be rewritten by taking the last term to the left-hand side as

$$-i\hbar \frac{\partial}{\partial x}(x\psi(x)) + i\hbar x \frac{\partial \psi(x)}{\partial x} = -i\hbar\psi(x),$$ (3.46)

or

$$\left[-i\hbar \frac{\partial}{\partial x} x + i\hbar x \frac{\partial}{\partial x} \right] \psi(x) = [px - xp]\psi(x) = -i\hbar\psi(x).$$ (3.47)

In the term on the right of the first equality, we have reintroduced the definition of the momentum (3.44). This combination of terms is called a *commutator bracket* and is written in shorthand as

$$[x, p] = [xp - px]. \tag{3.48}$$

This so-called commutator relationship arises between two operators and is different from zero (as the preceding case shows) for operators that do not *commute* (their order is important). The interpretation is that x is also an operator, but is a multiplicative operator rather than a differential operator, at least within this position representation.

The commutator relationship between two noncommuting operators was formulated by Heisenberg to be a central part of quantum theory. This is now known as the Heisenberg uncertainty relation. In the case of position and momentum, (3.46) yields

$$[x, p] = i\hbar. \tag{3.49}$$

(Recall that all of the terms in this equation still operate on the wave function.) The fact that these two operators represent dynamical quantities means in essence that it is impossible to measure them simultaneously, as discussed above. The Heisenberg uncertainty relation sets limits on the accuracy at a fundamental level. If we have $[A, B] = C$, then the uncertainty[‡] ΔA in determining the effects of operator A and the uncertainty ΔB in determining the effects of operator B are fundamentally limited to

$$\Delta A \Delta B \geqslant \frac{|C|}{2}. \tag{3.50}$$

In the case of momentum and position, this leads to

$$\Delta x \Delta p \geqslant \frac{\hbar}{2}. \tag{3.51}$$

[‡]By uncertainty in an operator, we refer to the mean-square deviation $\Delta A = \sqrt{\langle A^2 \rangle - \langle A \rangle^2}$, where the expectation is defined following (3.27) as $\langle A \rangle = \int_0^\infty A|\psi|^2 \, dx$.

Let us consider an example. Suppose we are able to localize a particle to within 10^{-11} m. Equation (3.51) says that we can determine the momentum only to within $\Delta p > 5 \times 10^{-24}$ kg-m/s or, for a free electron, $\Delta v > 5.8 \times 10^6$ m/s! The fact that the particle is highly localized, means that its wave packet is highly localized. However, we can say nothing about whether or not the particle is moving. We only can say that the velocity cannot be determined to better than Δv. If we knew its energy, we could determine its average velocity, but we don't know this without further information.

Many people often mistake the use of the uncertainty principle, and it is important to remember that it only applies to two noncommuting operators that represent the dynamics of the particle. It does not apply to the time. While there is a derivative with respect to time in the Schrödinger equation, time itself is not an operator. That is, in normal physics, classical as well as quantum, time is merely an indicator that we use to talk about *when* something happened. In classical (nonrelativistic) physics, the dynamical variables are x and p, and we talk about phase space that relates these two. In three dimensions, this phase space is a six-dimensional space and the dynamics is a trajectory through this space. Time is used to indicate some relationship on this trajectory to our subjective view of when an event started ($t = 0$ commonly), and when it was observed. The view of time is an argumentative subject for the philosophy of science. Here, we merely remark that there is no quantum mechanical uncertainty relationship involving time.

People often assume a commutator between energy and time, but this is a classical property that arises from difficulties in measurement. It is not a quantum result. However, we can illustrate the classical problem. If one considers a time-varying quantity with a simple exponential decay as $e^{-t/\tau}$, for $t > 0$, then the Fourier transform yields a functional form as

$$\frac{1}{1/\tau - i\omega}. \tag{3.52}$$

In Fig. 3.6, we plot the power spectral density (the magnitude squared of this equation). Clearly, if we want to accurately measure this time varying quantity, then we must use a measurement system

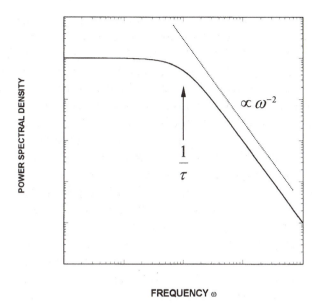

Figure 3.6. Power spectral density $P(\omega) = 1/[(1/\tau)^2 + \omega^2]$ of the exponentially decaying function $e^{-t/\tau}$. The spectrum is plotted on a logarithmic scale and a clear transition is resolved in this plot at frequency corresponding to $1/\tau$.

that can detect all important frequencies in the power spectrum of (3.52). This requires frequencies at least up to $1/\tau$. The factor τ itself is often described as the "uncertainty" or decay time of the exponential, and if we describe the $\Delta\omega$ required as the required range of frequencies from zero up to $1/\tau$, then we can write

$$\Delta\omega\tau > 1 \rightarrow \Delta E\tau > \hbar. \tag{3.53}$$

While $\hbar$ appears here, it only arrives through the Planck relation in what is otherwise a completely classical result (in the field of signal processing, this result is known by the name of Shannon's theorem, and in other fields as the Nyquist theorem) of Fourier transforms. This is a measurement problem and not a quantum mechanical result of the Heisenberg uncertainty relation. Nevertheless, it is often encountered in quantum mechanics, not the least because it is an important quantity—just not a quantum mechanical one. An example of the use of (3.53) lies in connecting the factor τ to the delay time of a circuit. Equation (3.53) tells us that if we want to have 1 GHz response within the circuit, then the delay time must be less than $1/2\pi$ ns, or less than 0.16 ns. Another example, often found in treating quantum systems, is for the decay of a quantum state. If the occupation probability of the state decays with a lifetime of 1 ns, then we cannot determine the energy of this state to an accuracy better than 6.6×10^{-7} eV, no matter how long we take to make the measurement (provided of course that we measure for longer than a nanosecond).

3.4. SOME SIMPLE POTENTIALS

We have seen in the previous paragraphs that when we are dealing with wavelike phenomena, the classical equations of motion are not easily applicable, and we must use the Schrödinger equation. It is important to note that the *forces* acting on the particle no longer appear explicitly in this new equation. Instead, they are represented in the term $V(x)$, which is the potential energy. It is therefore necessary to express the force in terms of a potential as

$$F = -\frac{\partial V(x)}{\partial x} \qquad (3.54)$$

in one dimension. [It is important to note that $V(x)$ here is the potential *energy* and not the electrostatic potential — here, we will try to write the electrostatic potential as $\phi(x)$ with $V(x) = -e\phi(x)$, where e is the negative charge on an electron, 1.6×10^{-19} Coulomb. Thus, an electrostatic potential of 1 V yields a potential energy of -1 eV $= -1.6 \times 10^{-19}$ J.] Since only the potential energy, and not the force, appears in the Schrödinger equation (3.38), it appears that only a special class of conservative forces can be treated. These forces are those that arise from a potential energy. This is not actually true, since a kinetic energy term also appears. Other types of forces, such as magnetic forces, can be introduced through a modification of the spatial derivative term within the kinetic energy. However, the treatment of such forces is beyond the scope of this text. It is fortuitous that the applications of the Schrödinger equation of interest are those in which the forces arise from a potential energy only. Moreover, we will in general seek only spatial solutions for which the results of (3.41) are applicable.

Let us consider first the special case in which the potential energy is constant everywhere—$V(x) = V_0$. Then, we can rewrite (3.41) as

$$-\frac{\hbar^2}{2m}\frac{d^2\psi(x)}{dx^2} = (E - V_0)\psi(x). \tag{3.55}$$

From this, it is clear that the constant potential V_0 provides a reference level for the energy. The dynamics are measured from this reference level, and we note that it is always possible to take $V_0 = 0$, although we shall not do this at present. We can rewrite (3.55) by introducing the wavenumber as

$$\frac{d^2\psi(x)}{dx^2} + k^2\psi(x) = 0, \qquad k^2 = \frac{2m(E - V_0)}{\hbar^2}. \tag{3.56}$$

The second equation, on the right of (3.56), connects the wave number k to the energy E. So long as $E > V_0$, the total energy is larger than the potential energy, and therefore the kinetic energy is a positive quantity. For these conditions, (3.56) may readily be solved to find solutions as

$$\psi(x) = Ae^{ikx} + Be^{-ikx}. \tag{3.57}$$

The quantities A and B are constants to be decided by normalization. The first term on the right-hand side of (3.57) can be coupled with (3.43) to give a plane wave, which propagates in the positive x direction. The second term on the right-hand side of (3.57) can be coupled with (3.43) to give a plane wave, which propagates in the negative x direction. The problem with these simple plane waves is that their squared magnitude is independent of position, and is everywhere $= 1$. Hence, the normalization integral (3.29) can not be evaluated. As a result, a different normalization will be required. We will not deal with this right now, as the problem can be dealt with in a straightforward manner for the potentials of interest.

Now, what about the case where $E < V_0$? In this case, the total energy is less than the potential energy and the kinetic energy is negative. Clearly, this can not correspond to a propagating plane wave, or to a wave packet that moves as a particle. In fact, this is not a classical regime for particles at all. For this case, we rewrite (3.56) as

$$\frac{d^2\psi(x)}{dx^2} - \gamma^2\psi(x) = 0, \qquad \gamma^2 = \frac{2m(V_0 - E)}{\hbar^2}. \tag{3.58}$$

Now, the solutions to this equation can be written as

$$\psi(x) = Ce^{\gamma x} + De^{-\gamma x}. \tag{3.59}$$

These two solutions are damped exponentials representing non-propagating modes. Again, C and D are constants to be evaluated by normalization at some point.

To demonstrate how these solutions can be used, we consider a potential that varies in position, but is piecewise constant. That is, we define a potential as

$$V(x) = \begin{cases} 0, & x < 0, \\ V_0, & x \geqslant 0. \end{cases} \tag{3.60}$$

This potential is shown in Fig. 3.7. There are two possible cases to be considered. In the first, the energy is below V_0, so that plane waves, in the region $x < 0$, couple to damped exponentials, in the region $x > 0$. The second case corresponds to an energy above V_0 so that plane waves exist in both regions. Even in this latter case, however, a discontinuity exists since the wave numbers are different in the two regions. The problem is mainly one of deciding how to match the wave functions in the two regions on either side of $x = 0$.

If the potential energy is well behaved (has no delta function properties), then one can integrate the Schrödinger equation (3.41) across the interface at $x = 0$. We can assume that the wave function is well behaved and has no divergences in this region, so that one arrives at the result that the value of the derivative of the wave function must be continuous across the interface. A second integra-

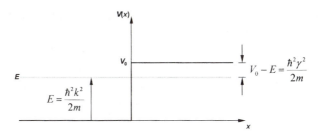

Figure 3.7. Schematic illustration of the different parameters used to describe the quantum mechanical transmission of particles through a potential step whose height V_0 is *greater* than the total energy E of the incident particle.

tion shows that the wave function itself must be continuous across the interface. These two conditions are sketched in Fig. 3.8, and set the necessary boundary conditions to solve the problem of the potential discontinuity.

We consider first the case for which $E < V_0$. It is assumed that the particle arrives from the far negative regions of position with a plane wave moving to the right. At the interface, it is reflected into a plane wave moving back to the left, while a small portion penetrates into the damping region (positive x). We can therefore write the wave function in the negative x region as

$$\psi_<(x) = Ae^{ikx} + Be^{-ikx}, \qquad k^2 = \frac{2mE}{\hbar^2}, \qquad x < 0. \qquad (3.61)$$

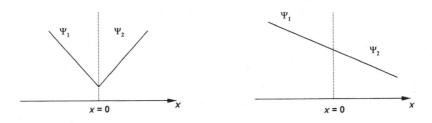

Figure 3.8. Matching the different wave function solutions at a potential boundary (indicated by the dotted lines). In the left-hand case, the different wave function solutions have the same value at the boundary but their slopes are discontinuous. This is physically unacceptable. In the right-hand case, however, the different wave function solutions have the same value *and* their slopes are continuous. This case represents a satisfactory solution of the boundary matching.

Here, A and B are the constants from (3.57). In the positive x region, we take only a single damped exponential. To use both terms of (3.59) would lead to a divergence at large x, which is unphysical. Thus, we write the wave function in the positive x region as

$$\psi_>(x) = De^{-\gamma x}, \qquad \gamma^2 = \frac{2m(V_0 - E)}{\hbar^2}, \qquad x > 0. \qquad (3.62)$$

The problem has now been reduced to using the boundary conditions to evaluate the constants A, B, and D. Our interest is in the reflection of the wave, and the resulting interference phenomena that arises between the reflected (represented by the term with coefficient B) and the incident waves (represented by the term with coefficient A). The values for k and γ are indicated in Fig. 3.7 with respect to the potential itself. To proceed, we impose first the continuity of the wave function at $x = 0$:

$$A + B = D. \qquad (3.63)$$

As a second condition, we impose the continuity of the derivative of the wave function at $x = 0$:

$$ik(A - B) = -\gamma D. \qquad (3.64)$$

Solving these last two equations together, we can find the values for B and D to be

$$D = \frac{2ik}{ik - \gamma} A, \qquad B = \frac{ik + \gamma}{2ik} D = \frac{ik + \gamma}{ik - \gamma} A. \qquad (3.65)$$

We note that the amplitude of the reflected wave function satisfies

$$\left| \frac{B}{A} \right|^2 = 1, \qquad B = e^{-2i\theta} A, \qquad \theta = \arctan\left(\frac{\gamma}{k} \right). \qquad (3.66)$$

This equation shows that the magnitude of the reflected wave is exactly equal to that of the incident wave. The reflection from the nonpropagating boundary produces only a phase shift in the reflected wave (relative to the incident wave). If we write the combined waves

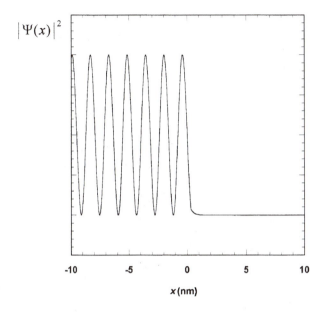

Figure 3.9. Squared amplitude of the wave function as a function of position in the region near a potential step. The step is located at $x = 0$ and has a height of 0.3 eV, whereas the total energy of the particle, which is incident from the left, is 0.15 eV.

as a single wave, we get

$$\psi_<(x) = A[e^{ikx} + e^{-i(kx + 2\theta)}] = Ae^{-i\theta}[e^{i(kx + \theta)} + e^{-i(kx + \theta)}]$$
$$= 2Ae^{-i\theta}\cos(kx + \theta). \tag{3.67}$$

In Fig. 3.9, we plot the squared magnitude of the wave function in the two regions. Clearly, the normalized squared magnitude of (3.67) is just $(A = 1)$

$$|\psi_<(x)|^2 = 4\cos^2(kx + \theta) = 2\{1 + \cos[2(kx + \theta)]\}. \tag{3.68}$$

The first zero of the total wave function occurs when the cosine is equal to -1, or

$$x_0 = \frac{1}{k}\left(\pm\frac{\pi}{4} - \theta\right). \tag{3.69}$$

The upper sign is used if $\theta > \pi/4$, and the lower sign is used if $\theta < \pi/4$. Successive zeroes are spaced by a quarter wavelength.

The transmitted wave has an amplitude that is also bounded. The squared magnitude of the transmitted wave is given by

$$|\psi_>(x)|^2 = A^2 \cos^2\theta e^{-2\gamma x}. \qquad (3.70)$$

This is also plotted in Fig. 3.9. There are several points of interest in (3.70) and (3.68). Both the amplitude of the transmitted wave and that of the reflected wave depend on both propagation variables k and γ. This is an example of a *nonlocal* effect. For large negative values of x, the wave function still depends on the properties of the region $x > 0$. In general, quantum effects are nonlocal. The amplitude of the wave function in the region $x > 0$ does not vanish unless the potential barrier is made infinitely large. In that case, $\gamma \rightarrow \infty$, and the wave is damped immediately to zero, $\psi_> = 0$. In addition, for this limiting case, the angle $\theta = \pi/2$, and $B = -A$.

Let us now turn to the second case, for which $E > V_0$. Now, the wave is propagating in both regions, as shown in Fig. 3.10, where the two wave vectors are defined as well. Again, we assume that a plane wave arrives from the large negative x region and propagates to the right. At $x = 0$, a portion is reflected and a portion is transmitted. Thus, we can write the wave function for the region $x < 0$ as

$$\psi_<(x) = Ae^{ikx} + Be^{-ikx}, \qquad k^2 = \frac{2mE}{\hbar^2}, \qquad x < 0. \qquad (3.71)$$

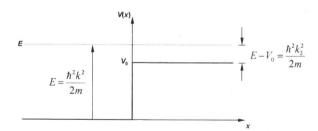

Figure 3.10. Schematic illustration of the different parameters used to describe the quantum mechanical transmission of particles through a potential step whose height V_0 is *less* than the total energy E of the incident particle.

This, of course, is just (3.61), as nothing has changed in the region for $x < 0$. For the region $x > 0$, however, we take only a plane wave propagating away from the interface (toward the right) as

$$\psi_>(x) = Ce^{ik_1 x}, \qquad k_1^2 = \frac{2m(E - V_0)}{\hbar^2}, \qquad x > 0. \qquad (3.72)$$

Once more, we have reduced the problem to one in which we only need to determine the constants A, B, and C. Again, we match the wave function and its derivative at $x = 0$ to give

$$A + B = C,$$

$$ik(A - B) = ik_1 C. \qquad (3.73)$$

These two equations can now be solved to yield

$$C = \frac{2k}{k + k_1} A, \qquad B = \frac{k - k_1}{2k} C = \frac{k - k_1}{k + k_1} A. \qquad (3.74)$$

Now, the amplitude of the reflected wave is no longer equal to the incident wave, and the wave function in the left region is given by

$$\psi_<(x) = A\left[e^{ikx} + \left(\frac{k - k_1}{k + k_1}\right) e^{-ikx} \right] = 2A\left[\cos(kx) - \frac{k_1}{k + k_1} e^{-ikx} \right].$$

$$(3.75)$$

The magnitude squared wave function becomes (we set $A = 1$ for convenience)

$$|\psi_<(x)|^2 = 1 + \left(\frac{k - k_1}{k + k_1}\right)^2 + 2\left(\frac{k - k_1}{k + k_1}\right) \cos(2kx). \qquad (3.76)$$

Once again, the amplitude of the total wave in this region can be larger than the incident wave due to the interference effects between the forward and reverse propagating waves. In fact, the values of

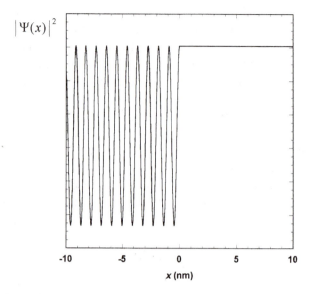

Figure 3.11. Squared amplitude of the wave function as a function of position in the region near a potential step. The step is located at $x = 0$ and has a height of 0.3 eV, whereas the total energy of the particle, which is incident from the left, is 0.45 eV.

(3.76) give a range for the square magnitude as

$$\left(\frac{2k_1}{k + k_1}\right)^2 \leq |\psi_<(x)|^2 \leq \left(\frac{2k}{k + k_1}\right)^2. \tag{3.77}$$

This wave is shown in Fig. 3.11. The transmitted wave now has the wave function

$$\psi_>(x) = A\frac{2k}{k + k_1}e^{ik_1x}, \tag{3.78}$$

and the squared magnitude is then ($A = 1$)

$$|\psi_>(x)|^2 = \left(\frac{2k}{k + k_1}\right)^2. \tag{3.79}$$

This is also shown in Fig. 3.11 for comparison. We note that the squared amplitude is a constant. Indeed, we have assumed that the transmitted wave is a single exponential, and this has a constant magnitude, with only a varying phase. Hence, we expect the squared magnitude to be a constant for $x > 0$. The discontinuity at $x = 0$ only determines the amplitude itself.

One interesting point from (3.79) is that, since $k_1 < k$, the amplitude of the transmitted wave is larger than 1. Just why this occurs is tied up with the de Broglie wavelength, which is $\lambda_1 = 2\pi/k_1$. The criteria that we introduced for the wave description of the particle is that the particle velocity is represented by the group velocity of the wave, described in (3.22). If we write (3.22) in terms of the de Broglie wave vector, then we arrive at

$$m v_{\text{group}} = m \frac{\partial \omega}{\partial k} = p = \frac{h}{\lambda} = \hbar k. \tag{3.80}$$

If we take the second term and the last term, we can introduce the Planck relation and write this as

$$\frac{m}{\hbar} \frac{\partial E}{\partial k} = \hbar k, \tag{3.81}$$

or, when the mass is independent of the wave vector k,

$$E = \frac{\hbar^2 k^2}{2m} + C = \frac{\hbar^2 k^2}{2m} + V(x). \tag{3.82}$$

In the last expression, it has been recognized that the constant of integration is just the k-independent potential energy. Thus, we can really connect the quantity $\hbar k$ as a momentum which is related to the particle momentum through the de Broglie wavelength relationship. The importance of this arises now with regard to the *flow* of probability—the probability current—from the input wave (arriving from the left) to the output wave across the interface. If the squared magnitude of the wave function is a probability density, then the product of this with the velocity is the probability *current*. (It would be a real current if we included the electronic charge.) We can define the transmission coefficient as

$$T = \frac{(\hbar k_1/m)|\psi_>(x)|^2}{(\hbar k/m)A^2} = \frac{k_1}{k}\left|\frac{C}{A}\right|^2 = \frac{4k_1 k}{(k + k_1)^2}. \qquad (3.83)$$

Since the incident probability has been taken to be unity, the reflection coefficient is

$$R = 1 - T = 1 - \frac{4kk_1}{(k + k_1)^2} = \left(\frac{k - k_1}{k + k_1}\right)^2 = \left|\frac{B}{A}\right|^2. \qquad (3.84)$$

If the energy is less than the potential height, the transmitted amplitude decays and the velocity is imaginary. Hence, $T = 0$ and $R = 1$.

We now can return to the question of why the amplitude is larger in the transmitted wave than in the incident wave. It is clear from (3.83) that the actual transmission coefficient is less than unity. If $k \gg k_1$, then $T \sim 4k_1/k \ll 1$. If the barrier is removed, then the transmission coefficient becomes unity. The smaller value of k_1 in the region $x > 0$ means that the velocity is much smaller here. The barrier creates a significant potential barrier, so that the kinetic energy is much reduced as the particle passes the point $x = 0$. Hence, probability density tends to accumulate in this region, leading to the larger amplitude of this quantity here. Yet, the probability density does not accumulate as rapidly as the velocity is reduced, so that the net transmission coefficient is less than unity. Let us consider an example. Suppose the barrier is 0.3 eV high, and a beam of free electrons is incident upon it with an energy ($x < 0$) of 0.4 eV. We then find that the velocity of the incoming particles is 2.65×10^7 cm/s, whereas that of the transmitted particles is 1.33×10^7 cm/s. The amplitude of the transmitted probability density, relative to that of the incident amplitude, is 1.78. However, the transmission coefficient is only 0.89. Hence, about 89% of the particles are transmitted and 11% are reflected at the interface. But, because the particles move slower in the region $x > 0$, their density increases as more particles arrive from the left than leave from the right. The incoming particles slow down, and build up the density, so that one needs more slower moving particles to carry the necessary current. Classically, all the particles would be transmitted, so that the reduction is a wave phenomena, which we attribute to the quantum effects of the particle as a wave.

3.5. TUNNELING THROUGH BARRIERS

In the previous section, we considered the electron wave impinging upon a potential barrier. The wave under the barrier decays exponentially, according to (3.62), at least when the energy is $E < V_0$. The decay distance is given by γ. Now, what happens if the barrier stops at a point d, which is smaller than a few times $1/\gamma$? A small fraction of the original wave then exists at this second boundary and can now excite another propagating wave, as we pass into the free region on the right-hand side of the barrier $(x > d)$! This is a more complicated situation, but the above treatment may readily be extended to one in which the barrier is of a finite width. Consider the case for which

$$V(x) = \begin{cases} 0, & x < 0 \\ V_0, & 0 \leqslant x \leqslant d \\ 0, & x > 0. \end{cases} \qquad (3.85)$$

Then the solutions for regions 1 $(x < 0)$ and 2 $(0 < x < d)$ are almost the same as before, but we must add a region 3 $(x > d)$. As before, we must also consider two cases that are given by the conditions that the incident energy is below the barrier or above the barrier. We begin with the former case, where $E < V_0$. The situation is shown in Fig. 3.12. The wave function in region 1 is now given by (3.61) as

$$\psi_1(x) = Ae^{ikx} + Be^{-ikx}, \qquad k^2 = \frac{2mE}{\hbar^2}, \qquad x < 0. \quad (3.86)$$

In region 2, we will need to have both wave solutions because it is possible for part of the wave to reflect from the second surface and propagate back to the left. Thus, we write the wave function for region 2 using (3.59) as

$$\psi_2(x) = Ce^{\gamma x} + De^{-\gamma x}, \qquad \gamma^2 = \frac{2m(V_0 - E)}{\hbar^2}, \qquad 0 \leqslant x \leqslant d. \quad (3.87)$$

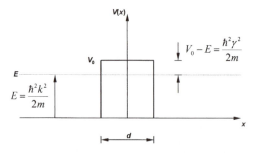

Figure 3.12. Schematic illustration of the different parameters used to describe the quantum mechanical transmission of particles through a potential barrier of finite width d whose height V_0 is *greater* than the total energy E of the incident particle.

In region 3, the wave function is similar to that of region 1, except that only the wave propagating to the right is retained. Thus, the wave function in region 3 can be written as

$$\psi_3(x) = Fe^{ikx}, \qquad k^2 = \frac{2mE}{\hbar^2}, \qquad x > d. \qquad (3.88)$$

As previously, the constants in (3.86)–(3.88) are evaluated by matching the wave function and its derivative at each of the two boundaries. At $x = 0$, this matching gives the two equations

$$A + B = C + D,$$
$$ik(A - B) = \gamma(C - D). \qquad (3.89)$$

Similarly, the matching conditions at the boundary $x = d$ lead to

$$Ce^{\gamma d} + De^{-\gamma d} = Fe^{ikd},$$
$$\gamma(Ce^{\gamma d} - De^{-\gamma d}) = ikFe^{ikd}. \qquad (3.90)$$

What we are really interested in is the fraction of the incident wave that makes it to the region beyond the barrier (region 3). Thus, the important constants are F and B. So, our approach is to solve (3.90) for C and D and use these in (3.89) to find F and B in terms of A.

The first task yields, from (3.90),

$$C = \frac{\gamma + ik}{2\gamma} F e^{ikd - \gamma d}, \qquad D = \frac{\gamma - ik}{2\gamma} F e^{ikd + \gamma d}. \qquad (3.91)$$

At the same time, we can obtain from (3.89) the intermediate step

$$A = \frac{ik + \gamma}{2ik} C + \frac{ik - \gamma}{2ik} D, \qquad B = \frac{ik - \gamma}{2ik} C + \frac{ik + \gamma}{2ik} D. \quad (3.92)$$

The two equations of (3.91) are inserted into the first of equations (3.92) to give

$$A = F e^{ikd} \left[\cosh(\gamma d) + \frac{k^2 - \gamma^2}{2ik\gamma} \sinh(\gamma d) \right], \qquad (3.93)$$

or

$$\frac{F}{A} = \frac{e^{-ikd}}{\left[\cosh(\gamma d) + \dfrac{k^2 - \gamma^2}{2ik\gamma} \sinh(\gamma d) \right]}. \qquad (3.94)$$

We can now perform a similar operation on the second equation of (3.92) to give

$$B = -F e^{ikd} \frac{k^2 - \gamma^2}{2ik\gamma} \sinh(\gamma d), \qquad (3.95)$$

and

$$\frac{B}{A} = -\frac{k^2 - \gamma^2}{2ik\gamma} \frac{\sinh(\gamma d)}{\left[\cosh(\gamma d) + \dfrac{k^2 - \gamma^2}{2ik\gamma} \sinh(\gamma d) \right]}. \qquad (3.96)$$

As mentioned, our interest is in the fraction of the incident wave that is transmitted to region 3. This is given directly by the squared magnitude of (3.94) since the velocity in regions 1 and 3 is the same

(same values for the wave vector k). Thus, we find that

$$
T = \left|\frac{F}{A}\right|^2 = \frac{1}{\left[\cosh^2(\gamma d) + \left(\frac{k^2 - \gamma^2}{2k\gamma}\right)^2 \sinh^2(\gamma d)\right]}
$$

$$
= \frac{1}{1 + \left(\frac{k^2 + \gamma^2}{2k\gamma}\right)^2 \sinh^2(\gamma d)}. \tag{3.97}
$$

In the last line, the $\cosh^2$ has been expanded as $1 + \sinh^2$ and then the two terms have been rationalized. The transmission coefficient T is termed the tunneling coefficient for this problem, as it is interpreted that the *incident particle has tunneled through the barrier*. There are two limits to (3.97) that are of interest. The first is for a large barrier, where $\gamma \gg k$. Then, we can expand the sinh function in terms of the dominant exponential and we approximate (3.97) as

$$
T \sim \frac{k^2}{\gamma^2} e^{-2\gamma d}, \qquad \gamma \gg k. \tag{3.98}
$$

In this case, the tunneling coefficient decreases exponentially with the thickness of the barrier, and many people ignore the prefactor as well. On the other hand, when the decay term is small, we need to carefully expand the sinh function to pass to the limit of small tunneling. In this limit, $\sinh(x) \sim x$, so that (3.97) becomes

$$
T \sim \frac{1}{1 + (kd/2)^2}, \qquad \gamma d \ll 1. \tag{3.99}
$$

In Fig. 3.13, we plot the tunneling coefficient as a function of energy, where the barrier height is assumed to be 0.3 eV, and the particle is a free electron. The barrier thickness is taken to be 1 nm. On the logarithmic plot, it can be seen that the tunneling coefficient is exponential for small T, and does not become unity for an energy at the top of the barrier. In fact, for these parameters, $k \sim 2.8 \times 10^9$ m^{-1}, so that $kd/2 = 1.4$. Hence, when the energy is at the top of the 0.3 eV barrier (of width 1 nm), the tunneling coefficient is only ~ 0.34. On the other hand, if we reduce the energy to 0.15 eV, then k and γ are equal in magnitude, and the tunneling coefficient is 0.073. Suppose we increase the barrier thickness by a factor of 2, to 2 nm.

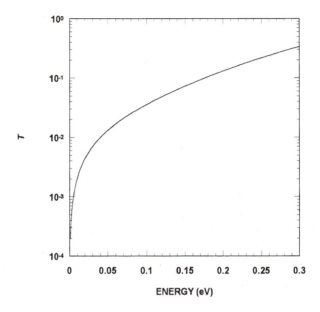

Figure 3.13. Quantum mechanical transmission probability for tunneling through a barrier of height 0.3 eV and width 1 nm. The variation of the tunneling probability is plotted over a range of incident particle energy from 0 to 0.3 eV. Note also that the variation of the transmission probability is plotted on a logarithmic scale.

Then, the tunneling rate at the top of the barrier ($E = 0.3$ eV) is reduced to 0.11, while that at 0.15 eV is reduced to 1.45×10^{-3}. Thus, increasing the barrier thickness by only a factor of 2 has reduced the tunneling coefficient by only a factor 3 at the top of the barrier — however, it has reduced the tunneling coefficient at 0.15 eV by almost a factor of 50!

Let us now turn to the situation in which the electron is incident with an energy that is larger than the barrier, as shown in Fig. 3.14. Classically, we would expect no effect of the barrier. As with the earlier barrier case, however, here we will still find a that the barrier has an effect. The only region in which the wave function changes is in region 2, where now the wave function is written as

$$\psi_2(x) = Ce^{ik_1x} + De^{-ik_1x}, \qquad k_1^2 = \frac{2m(E - V_0)}{\hbar^2}, \qquad 0 \leqslant x \leqslant d.$$

$$(3.100)$$

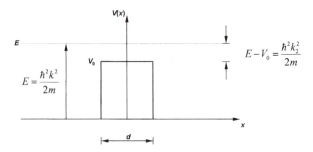

Figure 3.14. Schematic illustration of the different parameters used to describe the quantum mechanical transmission of particles through a potential barrier of finite width d whose height V_0 is *less* than the total energy E of the incident particle.

Now, matching the wave functions, and their derivatives, at the two interfaces yields the equations

$$A + B = C + D,$$

$$ik(A - B) = ik_1(C - D),$$

$$Ce^{ik_1d} + De^{-ik_1d} = Fe^{ikd},$$ (3.101)

$$ik_1(Ce^{ik_1d} - De^{-ik_1d}) = ikFe^{ikd}.$$

If we compare (3.101) with (3.89) and (3.90), we then find that the equations are precisely the same if we replace γ in (3.89) and (3.90) by ik_1. Since this is just a constant parameter, we can immediately use the previous solutions by making this same interchange. Noting that $\sinh(ik_1d) = i\sin(k_1d)$ and $\cosh(ik_1d) = \cos(k_1d)$, the tunneling factor (3.96) becomes

$$\frac{F}{A} = \frac{e^{-ikd}}{\left[\cos(k_1d) + \dfrac{k^2 + k_1^2}{2ikk_1}\sinh(k_1d)\right]}.$$ (3.102)

Now, the tunneling coefficient is found to be

$$T = \left|\frac{F}{A}\right|^2 = \frac{1}{\left[\cosh^2(\gamma d) + \left(\frac{k^2 + k_1^2}{2kk_1}\right)^2 \sinh^2(k_1 d)\right]}$$

$$= \frac{1}{1 + \left(\frac{k^2 - k_1^2}{2kk_1}\right)^2 \sinh^2(k_1 d)}.$$

(3.103)

Again, the $\cos^2$ term has been replaced by $1 - \sin^2$ and the two terms rationalized. Notice that (3.103) shows oscillatory behavior as a function of the incident particle energy. In fact, unity transmission is only obtained for $k_1 d = n\pi$, where n is an integer. At the top of the barrier, where k_1 approaches zero, (3.103) gives the same limit as (3.99). Now, however, for large k, the tunneling coefficient approaches unity, but with weak oscillations. The first point, for which the transmission becomes unity, is given by $k_1 d = \pi$. For the 1 nm barrier of height 0.3 eV, discussed above, we require $k_1 = \pi \times 10^9 \, \mathrm{m}^{-1}$, which is an incident energy of ~ 0.677 eV. On the other hand, if we increase the thickness to 2 nm, the required value of k_1 is $\pi/2 \times 10^9 \, \mathrm{m}^{-1}$, which corresponds to an incident energy of 0.309 eV. A thicker barrier leads to more rapid oscillations, with the first unity peak occurring at lower energy. The total tunneling probability T is plotted in Fig. 3.15 for a 2 nm barrier of 0.3 eV.

One of the most interesting applications of tunneling is the scanning tunneling microscope. In Fig. 3.16(a), the concept is sketched. A sharp metal tip is brought quite near a conducting surface. When a bias is applied between the tip and the surface (usually via a back contact), the electrons can tunnel from the tip, through the air gap, to the conducting surface. By monitoring the current as the tip is scanned across the surface, the distance can be determined from the current through (3.97); that is, the parameters for the tunneling current are all known except for the distance. Monitoring the current enables us to determine the distance from the exponential dependence of the current on this quantity. In practice, a servocontrol loop is used to move the tip closer to, or farther from,

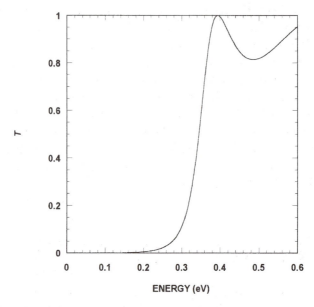

Figure 3.15. Quantum mechanical transmission probability for a barrier of height 0.3 eV and width 1 nm. The variation of the tunneling probability is plotted over a range of incident particle energy from 0 to 0.6 eV. Note that in this case, and in contrast to Fig. 3.13, the variation of the transmission probability is plotted on a linear rather than a logarithmic scale.

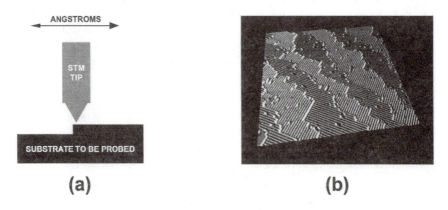

Figure 3.16. The scanning tunneling microscope (or STM). (a) Schematic diagram illustrating the basic principle of operation of the scanning tunneling microscope. The current that tunnels between the tip and the substrate is exponentially sensitive to the distance between them. (b) STM image of the silicon surface, illustrating the resolution of individual silicon atoms. *Picture courtesy of J. Lyding, University of Illinois.*

the surface to keep the current constant. By monitoring the height of the tip (from the voltage on the servo that moves the tip), the height of the surface is directly determined. The scanning tunneling microscope was invented by G. Binnig and H. Rohrer,[8] for which they shared a Nobel prize. The basic concept has been extended to the atomic force microscope in which force (or deflection of the point due to the force), rather than current, is measured. In Fig. 3.16(b), the (100) surface of silicon is shown. This image was obtained by atomic force microscopy at the University of Illinois. The two basis atoms of each lattice point are easily observed, as is the step where a new atomic bilayer has begun to form.

3.6. QUANTUM WELLS

In the previous section, the potential was a barrier that obstructed the incident wave if the energy were below the barrier height $(E < V_0)$, and to a lesser extent even when the energy was above the barrier height. We now want to consider the case in which the barrier is given a negative value, as depicted in Fig. 3.17. Here, the potential is given the value $-V_0$. However, the results are quite similar to those of the previous section. The wave functions in regions 1 $(x < 0)$ and 3 $(x > d)$ remain those of (3.86) and (3.88). In addition, the wave

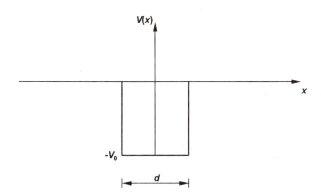

Figure 3.17. Schematic illustration of the different parameters used to describe the quantum mechanical transmission of particles through a potential well of finite width d and depth $-V_0$.

function of region 2 $(0 < x < d)$ remains that of (3.100) with the exception that the wave number is now given by

$$k_1^2 = \frac{2m(E + V_0)}{\hbar^2}. \tag{3.104}$$

Since the wave functions all have the same form, the solution remains that of (3.103). In (3.102), we can simplify the prefactor of the sinusoidal term as

$$|k^2 - k_1^2| = \left| \frac{2mE}{\hbar^2} - \frac{2m(E + V_0)}{\hbar^2} \right| = \frac{2mV_0}{\hbar^2}, \tag{3.105}$$

so that

$$\left| \frac{k^2 - k_1^2}{2kk_1} \right| = \frac{V_0}{2\sqrt{E(E + V_0)}}. \tag{3.106}$$

Thus, it is clear that this prefactor decays rapidly with the energy. In Fig. 3.18, we plot the transmission as a function of the energy for this potential case; a value of $V_0 = 0.3$ eV is used.

If we now want to study the case where the energy is below 0 (a state in the well), we have some difficulty, as we can no longer assume a propagating plane wave incident from the left. The waves are damped outside the well, so we must begin with a new assumption. Now, we shall assume that the potential is zero in the well, and nonzero outside the well, as shown in Fig. 3.19. To simplify the solution, however, we let the potential be infinitely large, with

$$V(x) = \begin{cases} \to \infty, & x < 0, \\ 0, & 0 < x < d, \\ \to \infty, & x > d. \end{cases} \tag{3.107}$$

With the potential diverging, we know from (3.59) that the wave is exponentially damped in an infinitely short distance. Thus, we take

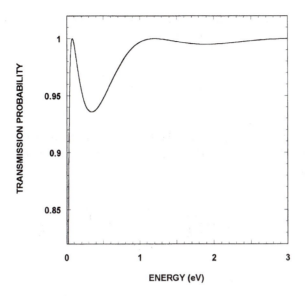

Figure 3.18. Quantum mechanical transmission probability for a potential well of depth 0.3 eV and width 1 nm. The variation of the tunneling probability is plotted over a range of incident particle energy from 0 to 0.3 eV. Note that the variation of the transmisssion probability is plotted on a linear scale.

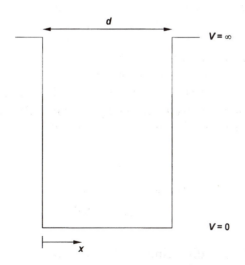

Figure 3.19. Schematic illustration of a potential well of infinite depth ($V_0 = \infty$).

the wave function to satisfy

$$\psi(x) = 0 \quad \text{for } x < 0, x > d. \tag{3.108}$$

That is, we will require that the wave function vanish at the edges of the quantum well. Inside the well, we now take the wave function from (3.57) to be

$$\psi(x) = Ae^{ikx} + Be^{-ikx}, \qquad k^2 = \frac{2mE}{\hbar^2}. \tag{3.109}$$

Requiring the wave function to vanish at $x = 0$ means that

$$A + B = 0, \qquad B = -A. \tag{3.110}$$

Requiring the wave function to vanish at $x = d$ means that

$$Ae^{ikd} - Ae^{-ikd} = 2A\sin(kd) = 0. \tag{3.111}$$

A choice can now be made. We can let $A = 0$, or work with the sin function. The former choice is not allowed, as the wave function then would be identically zero everywhere, which is not physical. Hence, we must set

$$\sin(kd) = 0 \rightarrow k = \frac{n\pi}{d}. \tag{3.112}$$

The value of k must take only certain numbers so that an integer number of half-wavelengths fit into the width d ($\lambda = 2\pi/k = 2d/n$, $n\lambda/2 = d$). There are an infinite number of these modes (recall that the potential is infinitely high), each with a wave number given by one value of n in (3.11). Thus, we can write the wave function as

$$\psi(x) = 2A_n \sin\left(\frac{n\pi x}{d}\right), \qquad 0 < x < d, \tag{3.113}$$

with

$$E_n = \frac{\hbar^2 k_n^2}{2m} = \frac{n^2 \pi^2 \hbar^2}{2md^2}, \qquad n = 1, 2, 3, \ldots. \qquad (3.114)$$

The coefficient A_n is now determined from the normalization condition (3.29), as

$$1 = 4A_n^2 \int_0^d \sin^2 \left(\frac{n\pi x}{d} \right) dx = 2A_n^2 \int_0^d \left[1 - \cos \left(\frac{2n\pi x}{d} \right) \right] dx = 2A_n^2 d.$$

$$(3.115)$$

We can now find the constant in (3.113) as

$$2A_n = \sqrt{\frac{2}{d}}. \qquad (3.116)$$

The wave function is now fully specified. Each value of n gives a different wave function and a different value of the energy. The energy *levels* are spaced by an increasing large amount, as the energy varies quadratically with n. The wave functions are called the eigenfunctions for the well, and the energies are the eigenvalues (*eigen* is a German word meaning single). In Fig. 3.20, the first few eigenfunctions for the well are plotted.

As an example, let us consider a quantum well that is 10 nm wide. We use the free electron mass, so the first energy level is given by (3.114) with $n = 1$. This is 3.77 meV (1 meV $= 10^{-3}$ eV). The next higher levels are at 15.1 and 34 meV. If the well width is reduced from 10 nm to 5 nm, each energy level is pushed upwards by a factor of 4, since the energy goes quadratically as the inverse of the width.

A careful examination of (3.113) shows that the derivative of the wave function is discontinuous at $x = 0$ and at $x = d$, since it is nonzero at these points, and zero just past these points. Previously, we used the continuity of the derivative in finding the solution. Here, however, we have ignored it. The rationale is that the Schrödinger equation is a second-order equation, which means that two and only two constraints are required for the solution. In the last section, we

94

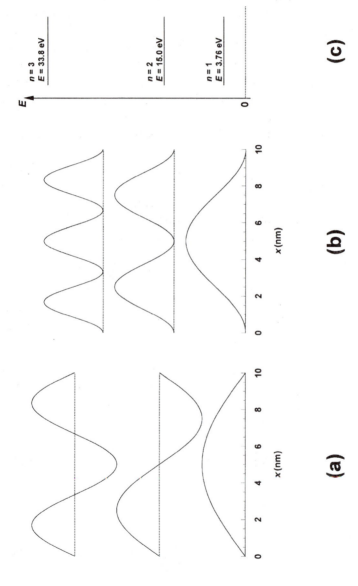

Figure 3.20. The first three eigenstates of a particle trapped in an infinitely deep potential well of length 10 nm. (a) The first three eigenfunction solutions $\psi_n(x)$, (b) the corresponding variation of the probability amplitude associated with the first three eigenstates, and (c) a schematic illustration of the first three energy levels.

used continuity of the wave function and its derivative. Now, we used two conditions on the wave function, one at each boundary, and this is all that is required. If we now try to specify the derivative, then the problem is overspecified. The discontinuity of the derivative is unavoidable. Note that this is, after all, consistent with the Schrödinger equation in the regions outside the well. Since V is infinite, this implies an unbounded second derivative in those regions. Therefore it is mathematically evident that the first derivative should contain a discontinuity at the boundaries. Physically, the discontinuity of the derivative implies an impulsive acceleration. Any particle that reaches the boundary sees this rather large accelerating force. The acceleration is in a direction such that the particle is urged back into the well.

We also note that the wave function is composed of two parts, one propagating in the positive x direction, and one propagating in the negative x direction. Thus, the uncertainty in k is $\Delta k = 2k$, whereas, the average value of the momentum is zero. That is, the actual wave function is a standing wave and does not propagate at all. It is reasonable to take Δx as the well width d, so that (3.51) becomes

$$\Delta x \Delta p = d \cdot 2k_n \hbar = 2n\pi\hbar, \qquad n = 1, 2, 3, \ldots. \qquad (3.117)$$

Even at the lowest energy level, this uncertainty is much larger than the minimum that the Heisenberg uncertainty relation would give us. In actual fact, we should compute the average (expectation) value of the position and the square of the position, and determine the uncertainty in position from

$$(\Delta x)^2 = \langle x^2 \rangle - \langle x \rangle^2, \qquad (3.118)$$

where

$$\langle O \rangle = \int_{-\infty}^{\infty} \psi^*(x) O \psi(x)\, dx = \frac{2}{d} \int_0^d O \sin^2(n\pi x d)\, dx. \quad (3.119)$$

Here, O is any arbitrary operator. We leave this to the problems.

3.7. THE PARTICLE IN A BOX

In the previous sections, we have proceeded primarily in a single dimension, the position x along the x axis. In the real world, however, we are faced with three dimensions. Normally, we refer these to the three coordinates x, y, and z. Thus, there is a Schrödinger equation for each of these three axes, and the three are combined into a vector form of the Schrödinger equation. In essence, the only change is that the second derivative with respect to x that appears in, for example, (3.38) is replaced by the Laplacian operator

$$\nabla^2 = \frac{\partial^2}{\partial x^2} + \frac{\partial^2}{\partial y^2} + \frac{\partial^2}{\partial y^2}. \tag{3.120}$$

In the time-independent form of the Schrödinger equation, this leads to a form of the Poisson equation (inhomogeneous Laplace's equation), which can actually be separated into the separate coordinates in five different coordinate systems. The most common of these are the common rectangular coordinates, but the group of separable coordinates also include spherical coordinates and cylindrical coordinates. The former (spherical coordinates) are composed of the radial distance r, the angle θ from the z axis, and the azimuthal angle ϕ around the z axis in the (x, y) plane. The cylindrical coordinates utilize the radial distance in the (x, y) plane from the z axis, the z axis itself, and the azimuthal angle previously given. There are special circumstances in which each of these coordinate systems is preferred, but it is only in the treatment below of the hydrogen atom that we will be concerned about the choice of coordinate system. Needless to say, each separate equation as a part of the overall Schrödinger equation contributes an eigenvalue. These are now called eigenvalues, rather than the eigenenergy, as it is only the sum of these that gives the energy in the wave equation. Thus, for example, in rectangular coordinates, a three-dimensional quantum well will yield values for $k_{x,n}$, $k_{y,m}$, and $k_{z,r}$. Here, n, m, and r are integers, and the energy is given by

$$E_{nmr} = \frac{\pi^2 \hbar^2}{2m} \left(\frac{n^2}{d_x^2} + \frac{m^2}{d_y^2} + \frac{r^2}{d_z^2} \right), \qquad n, m, r = 1, 2, 3, \dots. \tag{3.121}$$

The various d_i are the width of the well in the appropriate direction. The key point is that the three-dimensional solutions give three eigenvalues to be determined. In the case of the rectangular quantum well, they each have the same form. This is often not the case, however, as we will see in the next section.

A second problem that we must face prior to proceeding is that of normalization of the plane waves incident on the potential, or used in a variety of other applications. As mentioned, we cannot normalize them over the infinite space upon which they are defined. This would force the amplitude to zero (integrating a finite quantity over an infinite dimension gives an infinite result). Instead, we choose to define the wave in a fixed distance L. Hence, we define the wave as

$$\psi_L(x) = Ae^{ikx}, \qquad 0 < x < L. \tag{3.122}$$

However, instead of forcing the wave function to vanish at each limit, we take a different approach. By integrating the squared magnitude over this distance, the normalization requirement is

$$\int_0^L |\psi_L(x)|^2 \, dx = A^2 L = 1, \tag{3.123}$$

or

$$A = \frac{1}{\sqrt{L}}. \tag{3.124}$$

Yet, we have not specified how the wave function at 0 or L is matched to the wave function in the regions adjacent to this. There are many ways in which to do this, but the most useful (particularly for the next chapter) is to specify that the wave function is periodic in L. Certainly, the plane wave is periodic, and we need only make sure that the distance L coincides with the longest wavelength of interest in the problem (it is a wavelength because we require periodicity rather than just the zero of the wave function, as was required for the preceding quantum well). This, in itself, sets a difficulty. In essence, any solution can now be expanded in a Fourier series with periodicity $2L$. The longest half-wavelength mode that can fit into this length sets the fundamental mode, and no variation that is slower than this can be effectively studied. We will see in Chapter 4 that using the critical size of L, especially in solids, introduces some interesting properties of the wave vector k, not the least of which arises from this lowest wave number

$$k_{min} = \frac{2\pi}{L}.$$
(3.125)

This leads to the allowed values for k to be

$$k_n = \frac{2n\pi}{L}, \qquad n = 1, 2, 3, \ldots.$$
(3.126)

In fact, this result is slightly different from that of the quantum well in (3.112), and this is because the boundary conditions are different. In most problems, this normalization will drop out, and the limit $L \to \infty$ can be readily taken. In three dimensions, the plane waves are normalized to a rectangular box, so that three-dimensional plane wave takes the form

$$\psi_L(x, y, z) = \frac{1}{L^{3/2}} e^{i\mathbf{k}\cdot\mathbf{r}}.$$
(3.127)

This form is known as *box normalization*.

3.8. ATOMIC ENERGY LEVELS

One of the major achievements of quantum mechanics was its ability to explain atomic structure. For many years prior to the onset of quantum theory, atomic structure had been probed by optical spectroscopy. It was known that there were several series of atomic absorption lines, none of which could be explained by classical mechanics. As remarked at the beginning of this chapter, Bohr could explain some of these spectra by making the assumption that electrons trapped in certain orbits would not radiate energy as they were accelerated around the atom.[5] Yet, he could not explain other aspects of the observed absorption lines for the atoms. It remained for the full development of quantum theory by Schrödinger and Heisenberg to explain the various series of optical spectra.

In actual fact, even with the new quantum mechanics it is difficult to calculate the energy spectra (and, therefore, the optical absorption lines) of anything other than the simplest atom — hydrogen. The reason for this lies in the complexity of a multielectron atom. In helium, for instance, the two electrons and one nucleus create the classic three-body problem, which is known to be essentially unsolvable. Thus, one must use approximate results with many assumptions.

The simplest case is the hydrogen atom. This consists of a single electron orbiting a central nucleus. The attractive force between the electron and the nucleus is the Coulomb force

$$F(r) = -\frac{e^2}{4\pi\varepsilon_0 r^2}, \qquad (3.128)$$

where the negative sign tells us that the force is inward (attractive). Conventionally, the nucleus is located at the origin of the coordinate system, with the electron orbiting around it at a radius r. This naturally suggests spherical coordinates, and this will be the one discussed here. This *central* force leads to the potential

$$V(r) = -\int_\infty^r \frac{e^2}{4\pi\varepsilon_0 r^2}\, dr = -\frac{e^2}{4\pi\varepsilon_0 r}. \qquad (3.129)$$

This Coulomb force keeps the electron bound to the nucleus. In general, one now inserts this potential into the Schrödinger equation and solves for the energy levels. The details of this are beyond our current discussion (the hydrogen atom is solved in detail in Appendix A). Here, we discuss the major results, and consider what these tell us about the general atom. As we discussed in the previous section, solution of the Schrödinger equation in one dimension leads to one eigenvalue (or a set of eigenvalues). In the hydrogen atom, or in any spherically symmetric potential, this eigenvalue corresponds to the energy E. In three dimensions, there will be three eigenvalues. One of these corresponds to the solution in the dominant coordinate—the radial direction in the hydrogen atom. This will still determine the energy E, and is given the index n, just as in the quantum well. The other two dimensions correspond to the polar angle θ and the azimuthal angle ϕ. These correspond to orbital motion around the nucleus, which in classical mechanics is the angular momentum. This is also true in quantum mechanics, except this angular momentum is also quantized. We associate the variation in angle θ with the quantum number l, and the variation in angle ϕ with the quantum number m_z (not to be confused with the mass).

The radial quantum number determines the energy of the different levels, with this being given by

$$E_n = -\frac{1}{n^2}\left(\frac{m_0 e^4}{32\pi^2 \varepsilon_0^2 \hbar^2}\right) = -\frac{1}{n^2}\frac{\hbar^2}{2m_0 a_B^2}, \qquad n = 1, 2, 3, \ldots . \quad (3.130)$$

Here, we have indicated the mass by m_0 to distinguish it from the eigenvalue m (this will be done in this section only). In (3.130), the energy has been written in the form that it appears in the quantum well, although the variation with n differs. To do this, the *Bohr radius* is defined to be

$$a_B = \frac{4\pi\varepsilon_0 \hbar^2}{m_0 e^2}. \quad (3.131)$$

The Bohr radius has the value 0.053 nm for the free electron. The lowest energy level can hold only 2 electrons and is spherically symmetric. That is, $l = m_z = 0$. The values for l can take on values from 0 to $n - 1$. These states traditionally are described by Legendre polynomials, and there are $2l + 1$ values of m_z associated with each l. This is certainly confusing, so a little order is called for.

The lowest energy level is the $n = 1$ level, and for this level $l = m_z = 0$, as mentioned. The value $l = 0$ is associated with spherically symmetric states, so these are labeled s states for convenience. Thus, the lowest energy level is the 1s level. The second energy level has $n = 2$. Now, l can take either 0 or 1. As before, the state for $l = 0$ is spherically symmetric, so is the 2s state. The $l = 1$ states are associated with angular momentum along one of the axes (and are called the orbital quantum numbers), and are denoted the p states ($2p$ here). In the $2p$ states, the eigenvalue m_z can take on values of -1, 0, or 1. These quantum numbers are called the magnetic quantum numbers, as their degeneracy can be lifted by a magnetic field. There are $2l + 1 = 3$ values of m_z. Hence, in the $1s$ state there is one combination of integers to describe the state, while in the $2s + 2p$ states, there are four combinations of integers. In the $n = 3$ level, we have another angular momentum level, which is denoted the d states ($l = 2$), and there are an additional 5 combinations of integers, which give us 9 total for this level. Hence, the degeneracy of the energy shells is $1, 4, 9, \ldots$. But, experimentally, the degeneracy is $2, 8, 18, \ldots$. That is, there is another quantum number that is missing from our three dimensional description, and this quantum number leads to a doubling of the degeneracy. Thus, each coordinate state can hold two electrons, and only two electrons. Why?

Experimental studies of the quantum electron showed that an additional angular momentum was present.[9,10] This remained a mystery until Paul A. M. Dirac showed that the Schrödinger equation could be modified to exhibit solutions in which this new angular momentum would be explicitly present.[11] Dirac was looking for a form of the quantum equation that would be useful in relativity studies; in the process he discovered a new angular momentum coordinate which is called the *spin* of the electron. Thus, each of the two electrons in a coordinate state can hold two electrons, provided they have opposite spin—these are commonly referred to as "spin up" and "spin down" states. The spin quantum number is then given as $s = \pm 1/2$. The fact that only two electrons, with opposite spin, can occupy any coordinate quantum state is now known as the Pauli exclusion principle. It applies only to particles with half-integer spin. Photons have integer spin (± 1 corresponding to the circular polarization), so do not obey the Pauli exclusion principle. Taking the exclusion principle into account, there are now $2n^2$ electrons in each level, or "shell" of the atom. By following this "hydrogen atom" rule, we can almost create the periodic table of the elements, but major complications arise for the higher number of electrons one uses.

If life were as simple as the above discussion of the hydrogen atom, then we would have established the entire periodic table. But it is not this easy. The problem arises from the multiple electrons. Consider an atom with Z electrons. The first two electrons see an attractive potential of

$$V(r) = -\frac{Ze^2}{4\pi\varepsilon_0 r}. \qquad (3.132)$$

In fact, *both* electrons see this same potential, since they have opposite spin, and we cannot tell which spin state is occupied first. (There is actually a small additional energy related to the spin, but we discuss this much later.) These two electrons fill the $1s$ level. The $2s$ electrons, however, see a "core" potential of the nucleus plus the two $1s$ electrons, as

$$V_2(r) = -\frac{Ze^2}{4\pi\varepsilon_0 r} + \sum_{i=1,2}\frac{e^2}{4\pi\varepsilon_0|\mathbf{r}-\mathbf{r}_i|}, \qquad (3.133)$$

and the sum runs over the positions of the two $1s$ electrons. Once these two electrons are in the $2s$ level, then the next electrons see a core potential of

$$V_2(r) = -\frac{Ze^2}{4\pi\varepsilon_0 r} + \sum_{i=1,2}\frac{e^2}{4\pi\varepsilon_0|\mathbf{r}-\mathbf{r}_i|} + \sum_{i=3,4}\frac{e^2}{4\pi\varepsilon_0|\mathbf{r}-\mathbf{r}_i|}, \qquad (3.134)$$

where the last sum now runs over the positions of the $2s$ electrons. It is clear that the $2p$ electrons will see a different potential (a reduced one) from the $2s$ electrons. The hydrogen atom model suggests that the $2p$ and $2s$ electrons have the same energy. In fact, they do not. The energy levels of the $2p$ and $2s$ electrons are split, with $E_{2p} > E_{2s}$. This continues with the $n = 3$ level. First, the $3s$ levels are filled. Then the $3p$ levels are filled, but they lie at a higher energy level. Then, finally the $3d$ levels are filled (as mentioned, the levels for $l = 2$ are termed the d levels). We should ask whether it is possible for the $3d$ energy levels to lie above the $4s$ levels. The answer is quite generally "yes." For K and Ca, the $4s$ levels are filled (K has a single $4s$ electron, while Ca has 2), while the $3d$ levels remain empty. These are atomic number 19 and 20. Then, for atomic number 21 through 28, the $3d$ levels are sequentially filled, since this level lies above the $4s$ level, but below the $4p$ level. Thus, the splitting of the energy shell according to the angular momentum leads to quite complicated

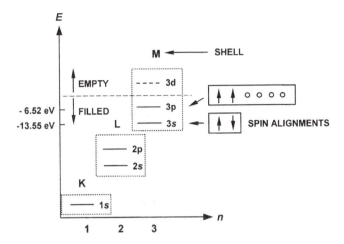

Figure 3.21. The energy-level structure of a silicon atom.

filling of the levels. Remember that the shell is designated only by the radial quantum number n. Hence, it is possible for an outer shell to begin filling before the next inner shell is completed.

Let us consider the case of the most common semiconductor—silicon. Silicon is atomic number 14, and so there are 14 electrons in the atom. Certainly, the $n = 1$ and $n = 2$ shells are filled. This accounts for 10 electrons. The last 4 electrons are the valence electrons, accounting for the chemical activity. Two of these go into the 3s level, while the last two go into the 3p level, which is not filled. The energy levels are shown in Fig. 3.21. We have also denoted the shells with another notation, which is often used by spectroscopists (where optical or x-ray). The $n = 1$ level is referred to as the K shell, while the $n = 2$ shell is referred to as the L shell. Similarly, the $n = 3$ shell is referred to as the M shell, and so on. We have also indicated the spin alignment in the 3s and 3p levels. In the K and L shells, all spin states are filled, so there is no net spin. This is also true in the 3s level. In the 3p level, however, a weak spin interaction (mentioned above) favors filling all of one spin type before beginning to fill the other. Thus, the two 3p electrons have the same spin state, shown as spin "up." This actually lowers the energy of the 3p level, and is known as "Hund's rule." Thus, there is a net spin angular momentum to the atom. In higher atomic number atoms, this spin angular momentum can be significant. Usually though, the spin angular momentum is randomly oriented. In some atoms, however, an interaction between the spins of one atom and those of an adjacent atom causes an ordering of the spins, which leads to a coherent orientation among neighboring atoms. The spin angular momentum leads to a magnetic moment, and such coherently oriented moments will lead to magnetization. We will discuss this in Chapter 8. Hund's rule is a statement that the spins will try to go into the levels in a manner that maximizes the angular momentum, and lowers the overall energy.

REFERENCES

1. A. A. Michelson, *Astron. Papers of U.S. Nautical Almanac Office* (U.S. Navy Department, Washington, D.C., 1882) Vol. I, Pt. III, 115; Proc. Am. Assoc. Adv. Sci. **28**, 124 (1879).
2. M. S. Longair, *Theoretical Concepts in Physics* (Cambridge, U.K.: The University Press, 1984).
3. M. Planck, Ann. Phys. **1(306)**, 730 (1900).
4. A. Einstein, Ann. Phys. **17**, 132 (1905).
5. N. Bohr, Phil. Mag. **26**, 1 (1913); **26**, 476 (1913).
6. A. Sommerfeld, Ann. Phys. **51**, 1 (1916).
7. L. de Broglie, Comp. Ren. Acad. Sci. Paris **177**, 507 (1923).
8. G. Binnig and H. Rohrer, Surf. Sci. **126**, 236 (1983).
9. G. E. Uhlenbeck and S. Goudsmit, Naturwiss. **13**, 953 (1925); Nature **117**, 264 (1926).
10. W. Pauli, Z. Phys. **43**, 601 (1927).
11. P. A. M. Dirac, Proc. R. Soc. (London) **A117**, 610 (1928).

PROBLEMS

1. Find the de Broglie wavelength for (i) an electron in free space with a velocity given by the thermal temperature of 300 K ($mv^2/2 = 3k_B T/2$), (ii) a helium atom with the same energy, and (iii) an α particle (He^4 nucleus) with energy of 25 MeV.

2. Graphically add three waves, of wave number 1.0, 1.1, and 1.2 cm^{-1}, which have equal amplitudes. Adjust all three of the waves to be in phase and at their positive peak at $x = 0$.

3. Suppose that an electron is confined to an infinite potential well of length 0.5 nm. What spectral frequencies will result from transitions between the lowest four energy levels?

4. Consider a two-dimensional potential in which

$$V(x, y) = \begin{cases} 0, & 0 < x < a, 0 < y < b, \\ V_0 \to \infty, & \text{elsewhere.} \end{cases}$$

What quantization conditions does this impose on the wave function (solve in two dimensions) and on the wave vectors k_x and k_y? How does the energy depend upon these quantities? Assume that the wave function must vanish at the edges of the potential well.

5. What geometric interpretation can be placed upon the length of a wave vector **k**, such that $k = |\mathbf{k}| = \sqrt{k_x^2 + k_y^2}$, for the two-dimensional system of problem 4. Describe a surface of constant energy in (k_x, k_y) space.

6. An electron microscope, operating at an accelerating voltage of 30 kV, illuminates a surface with energetic electrons. If the momentum is computed from the energy of the electrons, and a thermal spread of energy of 2500 K is assumed for the emitted electrons, what is the minimum resolution distance for this microscope?

7. In Maxwell's equations, one generally finds the variation in the current to be given by the continuity equation (in one dimension)

$$\frac{\partial J}{\partial x} = -e\frac{\partial n}{\partial t}.$$

Here, J is the current density and n is the particle density. If we use the probability density to represent the particle density, show that the current density may be written as

$$J(x) = \frac{\hbar e}{2mi}\left[\psi^*\frac{\partial \psi}{\partial x} - \frac{\partial \psi^*}{\partial x}\psi\right].$$

8. In the case of a finite tunneling barrier, such as given by (3.85), show that the current within the barrier is zero. How does this satisfy current continuity in the current through the barrier?

9. Solve the Schrödinger equation for a tunneling barrier, and compute the tunneling coefficient when the potential height is 2 eV and the incident energy is 0.5 eV. Take the barrier thickness as 0.2 nm. What if the barrier thickness is increased to 2 nm?

10. A particle confined to an infinite square well has an uncertainty in its position of the order of the well dimension, that is, $\Delta x \sim d$. The momentum must be at least as large as its uncertainty. From these facts, estimate the ground state ($n = 1$) energy level. Compare with the actual value.

11. If we take the wave functions for our infinite potential well, as given by (3.113), determine the expectation value for x and p, and the values of x^2 and p^2. Then, determine the uncertainty for each energy level according to (3.118) and (3.119), and the corresponding equation for the momentum.

12. If the potential barriers in the well problem are not infinite, then the four boundary conditions lead to a set of transcendental equations. Since the potential is not infinite, continuity of ψ, and $d\psi/dx$ is required at the boundaries. Obtain these transcendental equations and graphically solve them. Write a computer program to yield the values of the energy levels.

13. Using the solutions given for the hydrogen atom, compute the energy required for transitions between the $1s$ and $2s$ levels. What is the optical wavelength for this energy?

14. The ionization energy of hydrogen may be taken to be the energy of the $1s$ level (relative to the infinite distance, or vacuum, level). If a photon with wavelength of 15 nm is incident upon the atom, what is the kinetic energy of the emitted electron (the ionization energy is like the work function, but for the atom)?

15. The spherically symmetric solution of the Schrödinger equation for the $2s$ electron in the hydrogen atom can be written in the form

$$\psi(r) = C_{2s}(1 + C_1 r)e^{-rC_2/2},$$

where $C_2 = 1/a_B$ is the reciprocal of the Bohr radius. This wave function is often referred to as ψ_{200}. Why? Using the Schrödinger equation in spherical coordinates, determine C_1 from fitting this wave function to the equation. What is the energy level? What is the normalization constant C_{2s}? What is $\langle r \rangle$?

CHAPTER 4

Semiconductors

We found in the preceeding chapters that the atoms in a crystalline solid were located at the lattice points (and on the sites of the basis for that lattice). We also found the electrons were attracted to isolated atoms by the Coulomb potential, and the energy levels for these electrons were formed into shells. What happens when these electrons now sense the periodic nature of the lattice? Clearly electrons bound to atoms forms a different situation than free electrons zooming around in space. Thus, the nature of these electrons, confined to the atoms in the solid, is expected to be different. Yet, the electrons in a solid are different from the electrons in an isolated atom. Our task in this chapter is to bridge these two distinct regimes.

As we mentioned in Chapter 2, the atoms in nearly all semiconductors are covalently bonded. Covalent bonding involves the sharing of electrons among different atoms. In the hydrogen molecule (H_2), for example, the covalent bond contains two electrons. One electron is contributed from each atom of the molecule. Each of these two electrons moves back and forth between the two atoms (this is what the chemists call the resonating bond), and on average are situated between the atoms. In this special case, the two electrons come from the $1s$ level of the atoms.

If two silicon atoms were to come together, each would bring four electrons in the outer shell, and the two atoms could satisfy their desire for eight electrons by sharing, just as in the hydrogen molecule.

However, this is not what occurs, and the problem is the p states in the outer shell. The hydrogen molecule works because there is no angular momentum in the outer shell electrons. This is not the case in silicon, as there are two electrons in the $3s$ state and two electrons in the $3p$ state for each atom. When we bring the two silicon atoms together, the $3s$ state is already filled on each atom. These can't be shared unless one of the $3s$ electrons is "kicked" to a higher level (the $3p$ level). But this costs energy, and we usually release energy from the condensation into a solid. The solution is that the electrons are hybridized, which means that they form new hybrid wave functions which are combinations of $3s$ and $3p$ states.[1] Once this is done, the silicon atom can bond, for example, to four hydrogen atoms, forming the SiH_4 (silane) molecule. The four hydrogen atoms sit at the four corners of a regular tetrahedron (three-sided pyramid whose base is the same as any of the sides). Thus, these hybrids form a tetrahedral configuration, which can be seen in Fig. 4.1 (this sits in the diamond lattice, as can be seen by reference to Fig. 2.10). This tetrahedral coordination is common to nearly all semiconductors and is responsible for their similarities in properties. In the case of Fig. 4.1, however, the hydrogen atoms are replaced by four silicon atoms in the lattice itself. The difference between the diamond and zinc-blende structure arises from the fact that for GaAs, each Ga atom has four As atoms for neighbors, whereas each As atom has four Ga atoms

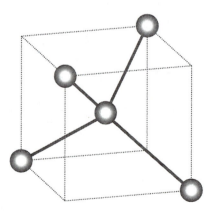

Figure 4.1. Illustration of the tetrahedral bonding arrangement found in the diamond and zinc-blende crystal structures. The solid lines denote chemical bonds between different atoms.

for nearest neighbors. (In essence, this is the lowest 1/8 of the unit cell shown in Fig. 2.10)

In this chapter, we examine how the tetrahedral coordination leads to the *energy bands* of the semiconductors. We then look at how the lattice vibrations occur, and how this affects the motion of the electrons. Here, we desire to arrive at the key factors, without delving too deeply into the detailed physics, as the latter can become quite complicated. Fortunately, we do not need the complicated details to understand the key effects that are important to us.

4.1. PERIODIC POTENTIALS

In the tetrahedral bonding arrangement, the four electrons from each Si atom are shared with the four nearest neighbor Si atoms (in a zinc-blende material, the atoms are different, but the sharing is the same). However, we still have the electrons localized in the atomic structure. How do we get the free carriers necessary to conduct electricity? The answer lies in the periodic nature of the lattice. We cannot use a free-electron model, as this neglects the interaction with the atomic potentials. (The free-electron model is for metals, where nearly all the electrons behave as if they were largely free from any atomic influence.) Free electrons provide a basis for which to develop the necessary modifications that lead us to the *band picture*. In Chapter 2, we discovered that x rays are diffracted by the atomic planes of the lattice. But electrons can also be viewed as waves. Not only does the electron sense the periodic potential of the atomic potentials, it also can be diffracted by the regular lattice planes of these atomic potentials. We determined that Bragg diffraction (2.8) would occur when the wavelength λ of the wave was related to the period of the atomic lattice a_0 through[2]

$$2a_0 \sin \theta = n\lambda, \tag{4.1}$$

or

$$k \sin \theta = \frac{n\pi}{a_0}, \tag{4.2}$$

where we have replaced the wavelength with $\lambda = 2\pi/k$ and rearranged the terms. Now, $k \sin \theta = k_x$, which is the primary single dimension with which we are working. Hence, diffraction occurs when

$$k_x = \pm \frac{n\pi}{a_0}, \tag{4.3}$$

and the minus sign has been inserted to account for the fact that n can vary over all integers, both positive and negative, but in (4.3) only positive integers are used. (The general vector $\mathbf{k}$ can be seen in the overlay to Fig. 2.16(b), where various directions were shown.) When the wave vector k_x reaches this condition, the electron wave will be diffracted. A similar condition will arise for each of the other directions.

The Bragg diffraction can reflect the electron back into the reverse direction. Hence, an incident wave propagating in the positive x direction will have a reflected component sent back into the negative x direction. But the crystal is symmetric; it is not possible to know which is the positive x direction and which is the negative x direction, as these are axes that we define in looking at the crystal. They are not intrinsic to the crystal. As a result, we expect that the waves propagating in the two directions will have equal amplitudes. When they interfere with each other, they produce standing waves. There are two types of standing waves that can arise from these two counterpropagating waves — they can interfere positively or negatively, and the resulting wave can be

$$\psi(x) = e^{ikx} \pm e^{-ikx}. \tag{4.4}$$

The upper sign produces a constructive interference, giving a wave as

$$\psi^{(+)}(x) = 2\cos\left(\frac{n\pi x}{a_0}\right). \tag{4.5}$$

At this point, we have to ask where the origin of the coordinates is to be placed. We have diffracted the electron wave from the atomic planes, but this occurs iregardless of where the origin is located. In fact, it is customary to locate the origin on the atomic position, so

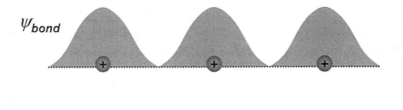

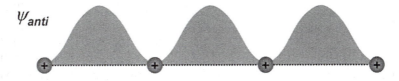

Figure 4.2. Schematic illustration of the bonding and antibonding wave functions for a simple diatomic molecule. The upper case shows the bonding orbital, for which the wave function is *nonzero* at the midpoint between the atoms. The lower figure shows the antibonding wave function. Note how this is *equal* to zero at the midpoint between the atoms.

that the lattice points in the x direction are at $na_0\mathbf{a}_x$, where $\mathbf{a}_x$ is a unit vector in the x direction. Then, the wave function (4.5) has its peak position on the atomic sites, so that the electron is tightly bound to the atomic core. This wave function is traditionally termed the *bonding* state. In Fig. 4.2, we illustrate the wave functions for the bonding and the antibonding states (described in the following).

The destructively interfering combination in (4.4) leads to a different wave function, which is given by

$$\psi^{(-)}(x) = 2i \sin\left(\frac{n\pi x}{a_0}\right). \tag{4.6}$$

This wave function has its probability density peaked between the atomic positions, and is asymmetric with respect to x. This asymmetry is appropriate for the electrons, and the wave is termed the *antibonding* wave function.

The bonding and antibonding wave functions lead to charge that accumulates at different locations within the unit cell of the lattice. The bonding wave function localizes charge on the atomic positions, whereas the antibonding charge localizes the charge between the atomic positions. The kinetic energy of the two waves is the same, since they have the same value of the wave vector k. However, the potential energy is different, since the charge associated with the wave function corresponds to different values of the atomic potentials. Another way of thinking about this is that the bonding state is a cooperative interaction, which always lowers the energy. On the other hand, the antibonding state is a competitive interaction, which costs energy — thus raising the energy level for this state. The bonding wave function is localized near the atomic core, where the potential is quite negative. Thus, the potential energy for this wave function is relatively negative in value. On the other hand, the antibonding wave function is localized between the atomic cores, where the atomic potential has risen to a higher value than at the atomic site. Thus, this wave function has a higher potential energy than the bonding wave function. Thus, we are led to the conclusion that these two wave functions, in fact, have different energy eigenvalues, and the difference between them is related to the effect of the atomic potentials acting on the waves. In essence, the diffraction at values of k given by (4.3) causes gaps to open in the "free electron" spectrum. These gaps arise from the interaction with the atomic potentials, and separate the bonding and anti-bonding energy states.[3] We sketch these gaps on the free electron spectrum in Fig. 4.3. The size of the gap depends on the strength of the interaction between the "nearly free" electrons and the atomic core potentials.

Thus, the primary role of the crystal potential in this model is to give rise to energy gaps at particular values of the momentum, where the electron wave undergoes diffraction by the lattice. Within these gaps, the plane waves cannot exist, and we refer to the gaps as *forbidden* energy gaps. In Fig. 4.3, the dotted line represents the energy spectrum for the free electrons in space with no atomic structure. On the other hand, the solid lines are the allowed energy states within the crystal, so that the diffraction-induced gaps separate allowed energy ranges. These allowed energy ranges are termed *bands*. As a general rule, the higher in energy the electron lies, the less it feels the atomic potential. As a result, the gaps become smaller and the bands become wider as the energy increases.

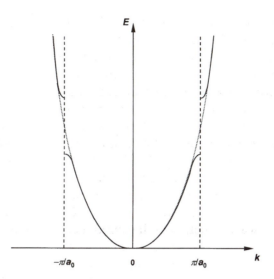

Figure 4.3. Electron energy spectrum, showing the formation of an energy gap due to diffraction of electrons by the crystal lattice. The dotted line denotes the free-electron dispersion $E = \hbar^2 k^2 / 2m$. The magnitude of the energy gap that opens in the actual electron spectrum at $k = \pm \pi / a_0$ depends on the strength of the interaction between the nearly free electrons and the atomic core potentials.

4.2. BLOCH'S THEOREM AND BRILLOUIN ZONES

The crystal lattice, and therefore the atomic potentials, have a certain periodicity that is a property of the material. This means that there exists a unit length a_0 for which the potential satisfies

$$V(x + a_0) = V(x). \tag{4.7}$$

We deal here with a one-dimensional case, but it is trivially extended to three dimensions. At the same time as we have this lattice periodicity, we also take into account that the entire structure is of some length $L = N a_0$, where N is the number of lattice sites in this chain. As with the box normalization of Chapter 3, we shall insist that the wave function is periodic over this length, so that

$$\psi(x + L) = \psi(x). \tag{4.8}$$

The wave function can vary from one atomic site to another, but the squared amplitude must be the same at each lattice site, otherwise one site would be preferred over another. Because the lattice is homogeneous from one site to another, we cannot have this happen, so the wave function can vary by at most a phase factor. Thus, the extension of (4.7) to the wave function gives[4]

$$\psi(x + a_0) = e^{i\varphi}\psi(x). \tag{4.9}$$

This equation is known as Bloch's theorem. If we repeat this process through the entire length of the crystal, we have

$$\psi(x + L) = e^{iN\varphi}\psi(x) = \psi(x). \tag{4.10}$$

Hence,

$$N\varphi = 2n\pi, \qquad \varphi = \frac{2n\pi}{N}. \tag{4.11}$$

If the wave functions are basically plane waves, then $\varphi = ka_0$, and

$$k = \frac{2n\pi}{Na_0} = \frac{2n\pi}{L}. \tag{4.12}$$

We can now see that n can take the values $0, 1, 2, \ldots, N$. That is, for *each* band, there are N independent values of n; for larger values, the phases begin to repeat and do not bring new information. By looking at Fig. 4.3, we see that each band region is defined by a set of values for k that lie in multiples of the range of (4.12). The lowest set is given by

$$-\frac{\pi}{a_0} < k \leqslant \frac{\pi}{a_0}. \tag{4.13}$$

This range of values for k defines what is called the *first Brillouin zone*. If we take any value of k, and reduce the phase by 2π, we are shifting k by the amount $2\pi/a_0$. This follows since $e^{i(k+2\pi/a_0)a_0} = e^{ika_0}e^{i2\pi} = e^{ika_0}$. Hence, this shift can be used to move all of the bands in Fig. 4.3 back into the minimum range defined by (4.13). This is shown in Fig. 4.4, and is termed the *reduced-zone* scheme. This is the way in which nearly all band structure is plotted, as we will see later in our further discussions of this concept.

The connection between the number of states in the band, which is $2N$ (taking into account the spin), and the number of electrons tells us a lot about the properties of the material. For example, the number of states ($2N$) is twice the number of atoms (or lattice sites, N) in the linear chain. A typical metal atom has only a single outer electron, so that only one-half of the states in the band is filled. On the other hand, the diamond lattice has 2 atoms per unit cell, so that

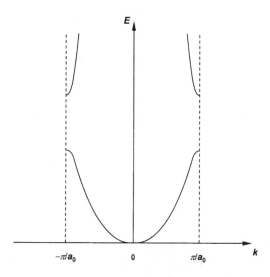

Figure 4.4. The nearly free electron spectrum shown in 4.2 is plotted in the *reduced-zone* scheme, by collapsing the energy bands into the first Brillouin zone.

there are $2 \times 2N$ electrons, where N is the number of lattice sites rather than the number of atoms, and this completely fills the band. Thus, there is a gap to the next band, and materials such as Si are not good conductors. Rather, they are more like insulators, except that the gap is not very large.

A crucial change has been made in the wave vector as well. Previously, the wave vector was clearly defined by its relationship to the electron's wavelength, and thus to the momentum of the particle. Now, however, the wave vector k is no longer a continuous variable. Rather, it has a well-defined discrete nature and can easily be changed to a larger or smaller value by the increment $2\pi/a_0$. In practice, this discretization is not easily observed, since there are likely to be more than 10^{23} values of k within each band. Yet if we are to still connect the quantity $\hbar k$ to the momentum of an equivalent particle, then great care must be exercised as to the range of k over which this can be done. In fact, we must restrain the range of k to lie within the first Brillouin zone. In this way, we will also have to carefully define the *mass* of the particle to which this wave vector refers. In this way, we refer to the wave vector k, within the restricted range of a single Brillouin zone, as the *crystal momentum*. Within this restricted range, we can identify it with the momentum of a free particle, but this must be done with care. We can easily see in Figs. 4.3 and 4.4 that the variation of the energy with wave vector does not follow the simple parabola defined by the free particle mass. This means that we must also change the properties that we assign to the equivalent free particle. In this regard, we must now determine what the *effective mass* is within the crystal. That is, in relating the momentum to a velocity of the equivalent particle, we can no longer use the mass of a free particle. Instead, it is necessary to find the equivalent mass that relates the particle properties (velocity) to the wave properties (wave vector). We return to this after we have obtained an analytic form for the energy bands.

4.3. THE KRONIG–PENNEY MODEL

Let us now consider a rather complicated model of the periodic potential and study how it gives rise to the bands in which we are

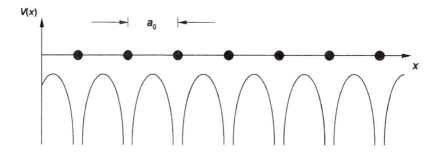

Figure 4.5. A one-dimensional array of atoms in a perfect crystal. In this figure, the atoms all reside at their equilibrium positions. Also shown is the corresponding crystal potential variation that the atoms give rise to. In a perfect crystal, this potential is a periodic function of position within the crystal.

interested.[5] In Fig. 4.5, we illustrate an idealized set of atomic potentials. Here, each atom sits in a region of *zero* potential, while potential barriers exist between these atoms. Such an array of potentials is called a *periodic* potential. Note that the width of each potential well, in which the atoms are situated, is a_0, while the periodic distance is given by this value plus the width of the barriers, d. (We take the potential amplitudes as V, although this will be allowed to become infinitely large at a later stage, at which point d will pass to zero. This is shown in Fig. 4.6.) Within the well region, we describe the wave function through the Schrödinger equation

$$\frac{d^2\psi}{dx^2} + \alpha^2\psi = 0, \qquad \alpha^2 = \frac{2mE}{\hbar^2}, \qquad 0 \leqslant x \leqslant a_0. \qquad (4.14)$$

Note that here we are using α for the propagation wave vector within the well, as the value k will be reserved for the coherent wave that exists within the entire structure. This will become clear shortly. Within the potential, we follow the same prescription as in Chapter 3 and write the Schrödinger equation as

$$\frac{d^2\psi}{dx^2} - \gamma^2\psi = 0, \qquad \gamma^2 = \frac{2m(V-E)}{\hbar^2}, \qquad -d \leqslant x \leqslant 0, \quad (4.15)$$

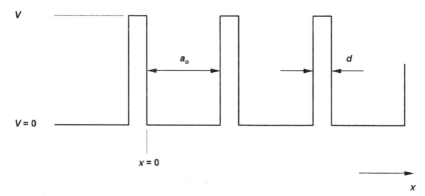

Figure 4.6. The form of the periodic potential that is assumed in the description of the Kronig–Penney model. Each of the potential wells is assumed to be generated by a particular atom within the crystal. The periodic potential variation shown here is a crude representation of that shown in Fig. 4.5.

Now, we require our overall wave function to have propagating waves with wave vector k; that is, we write the overall wave function in the form

$$\psi(x) = e^{ikx} u_i(x), \tag{4.16}$$

where the functions u_i have the periodicity of the lattice

$$u_i(x + d + a_0) = u_i(x). \tag{4.17}$$

The cell periodic functions u_i are the quantities we need to determine for the band structure. Using (4.17), we can rewrite the Schrödinger equation in the well region as

$$\frac{d^2 u_1}{dx^2} + 2ik\frac{du_1}{dx} + (\alpha^2 - k^2)u_1 = 0, \qquad 0 \leqslant x \leqslant a_0, \tag{4.18}$$

and in the barrier regions as

$$\frac{d^2 u_2}{dx^2} + 2ik\frac{du_2}{dx} - (\gamma^2 + k^2)u_2 = 0, \qquad -d \leqslant x \leqslant 0. \tag{4.19}$$

As discussed, we are seeking wavelike solutions, so we assume that we can expand the cell periodic functions as

$$u_i = e^{\delta_i x} \tag{4.20}$$

so that our Schrödinger equations become

$$\delta_1^2 + 2ik\delta_1 + (\alpha^2 - k^2) = 0, \qquad 0 \leqslant x \leqslant a_0, \tag{4.21}$$

in the well, and

$$\delta_2^2 + 2ik\delta_2 - (\gamma^2 + k^2) = 0, \qquad -d \leqslant x \leqslant 0, \tag{4.22}$$

in the barrier. This gives us the solutions

$$
\begin{aligned}
u_1 &= Ae^{-ikx+i\alpha x} + Be^{-ikx-i\alpha x}, \\
u_2 &= Ce^{-ikx+\gamma x} + De^{-ikx-\gamma x}.
\end{aligned} \tag{4.23}
$$

The constants must now be determined by the use of the boundary conditions.

As in Chapter 3, the boundary conditions that must be employed are to ensure that both the wave function and its derivative are continuous at each edge of the well and the barrier. However, we can impose these boundary conditions at only two points, so that some ingenuity must be used to ensure that the periodicity of the potential is imposed on the wave function. Obviously, we impose the continuity of the wave function and its derivative at $x = 0$, which gives

$$u_1(0) = u_2(0), \qquad A + B = C + D, \tag{4.24}$$

and

$$\left.\frac{du_1}{dx}\right|_0 = \left.\frac{du_2}{dx}\right|_0, \qquad ik(\alpha-k)A - ik(\alpha+k) = -(ik-\gamma)C - (ik+\gamma)D. \tag{4.25}$$

The next boundary condition will take the values of u_1 at $x = a_0$, but the values of u_2 at $x = -d$, which is a lattice displacement back. This ensures that the periodic nature of the potential is imposed on the wave functions. This now gives us

$$u_1(a_0) = u_2(-d), \qquad Ae^{-i(k-\alpha)a_0} + Be^{-i(k+\alpha)a_0} = Ce^{(ik-\gamma)d} + De^{(ik+\gamma)d},$$

(4.26)

and

$$\left.\frac{du_1}{dx}\right|_{a_0} = \left.\frac{du_2}{dx}\right|_{-d}, \qquad \begin{aligned} & ik(\alpha-k)Ae^{-ik-\alpha)a_0} - ik(\alpha+k)e^{-i(k+\alpha)a_0} \\ & = -(ik-\gamma)Ce^{(ik-\gamma)d} - (ik+\gamma)De^{(ik+\gamma)d}. \end{aligned}$$ (4.27)

These four equations represent the solution matrix for the four coefficients. However, there are no forcing functions — once we form the matrix equation, the right-hand side is zero. This means that we have a coefficient matrix that multiplies a matrix of the constants, and this is equal to zero. Hence, the only solutions that can arise are for the case in which these equations more than specify the solution — the determinant of the matrix must vanish.

Only those values of k for which the determinant vanishes are allowed values for a propagating wave. This determinant can be expanded into the equation

$$\frac{\gamma^2 - a_0^2}{2\gamma a_0} \sinh(\gamma d) \sin(\alpha a_0) + \cosh(\gamma d) \cos(\alpha a_0) = \cos[k(a_0+d)]. \quad (4.28)$$

We now take the limit $V \to \infty$, $d \to$ in such a manner that Vd remains finite. In this limit, we can write

$$\cosh(\gamma d) \to 1, \qquad \frac{\sinh(\gamma d)}{\gamma d} \to 1, \qquad (4.29)$$

so that

$$\frac{\gamma^2 d}{2a_0} \sin(\alpha a_0) + \cos(\alpha a_0) = \cos[ka_0]. \qquad (4.30)$$

The left-hand side of this equation depends on the energy (we recall that the prefactor remains finite). The right-hand side, however,

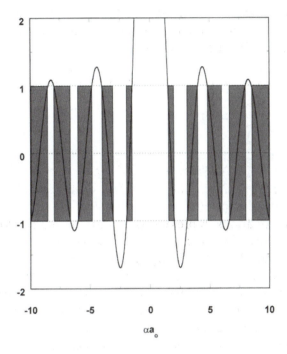

Figure 4.7. In this figure, the left-hand side of Eq. (4.30) is plotted as a function of the variable αa_0. Since the right-hand side of Eq. (4.30) can take values only between ± 1, we conclude that solution of this equation is only possible when αa_0 lies within certain allowed bands, which we illustrate as the shaded regions in Fig. 4.4. For values of αa_0 outside of these bands, solution of Eq. (4.30) is impossible.

depends only on the wave vector k that characterizes the wave function. Thus, there are only certain ranges of energy for which solutions are found to exist. In Fig. 4.7, we plot the left-hand side of (4.30) versus the value of αa_0. The right-hand side is limited to an absolute value of unity, so that the ranges of energy for which the left-hand side has amplitude less than unity correspond to allowed values of the energy. Within these regions, the right-hand side suggests that the energy bands are approximately cosinusoidal in nature.

The regions where the left-hand side has magnitude greater than unity correspond to regions for which no valid solution is available. This must correspond to the band gaps. These are referred to as forbidden energy regions. In actuality, these are regions for which our assumed form of the wave function (4.16) is not a valid solution. Other solutions might exist in these regions, but they do not correspond to our plane waves, which represent the motion of the particles. In the next section, we explore the nature of the co-sinusoidal bands further.

We note that Fig. 4.7 has left–right symmetry. This symmetry corresponds to the symmetry between $+k$ and $-k$. This symmetry is clearly evident in both Fig. 4.3 and Fig. 4.4. Consider the lowest band on each side of Fig. 4.7. The value of $\cos(ka_0)$ passes from $+1$ to -1. On the right-hand side of the figure, this occurs for positive k, while on the left-hand side of the figure, it occurs for negative k. These bands correspond to $0 < k < \pi/a_0$, and to $-\pi/a_0 < k < 0$, respectively. The two, taken together, represent the lowest band in either Fig. 4.3 or Fig. 4.4. Now, the second band is a little more challenging. We can interpret the limits in two ways. Consider the second band on the right-hand side of Fig. 4.7. Here, $\cos(ka_0)$ passes from -1 to $+1$. Yet we should interpret k as having larger values than those of the first band. Thus it is logical to take $\pi/a_0 < k < 2\pi/a_0$. The equivalent second band on the left-hand side of Fig. 4.7 gives the negative of these values. So these bands are the second bands in Fig. 4.3. However, we could as easily require that k always lie within the first Brillouin zone, in which case, the second band on the right would have values of k beginning at π/a_0 and running down to 0. The second band on the left would produce the negative of this range. Again, taken together, these two produce the second band in Fig. 4.4, the reduced zone scheme.

The results of the Kronig–Penney model are completely consistent with the bands and gaps that arise from the nearly free electron model. The bands are almost cosinusoidal in nature—they are distorted slightly by the first term on the left-hand side of (4.30). These one-dimensional models always produce nearly cosinusoidal bands, so that we can interpret the energy bands and gaps as being these nearly free electron bands, but with the gaps opened by the interaction of the particles with the atomic potentials. Moreover, we note from Fig. 4.7, that the gaps become smaller as we move to larger values of the energy (αa_0), and the bands must become increasingly wider in this same limit, as discussed earlier. In the next section, we take the opposite view of band formation—weak interactions as opposed to the strong ones used here, but arrive at the same conclusion.

4.4. NEAREST NEIGHBOR COUPLING—THE TIGHT-BINDING APPROACH

So far, we have forced the properties of the crystal—periodic potential with the atomic potentials situated on the lattice sites—on to the wave vector and energy of the electronic wave function. The potential leads to interactions, primarily at the values of wave vector appropriate for Bragg reflection of the waves, which causes gaps to open in the energy spectrum. At present, however, we have no exact, or approximate, form for the energy band within the allowed range of k that corresponds to a single (e.g., the first) Brillouin zone ($-\pi/a_0 < k < \pi/a_0$). Let us now turn our attention to how this can be determined. Consider the array of atoms that was shown in Fig. 4.5. Each of the atoms can be labeled with an integer, such that the relative position of the atom (from some zero reference point) may be defined by the value $x = ja_0$, where j is an integer. Each atomic site is spaced from the next one by the lattice constant a_0. Normally, each atomic energy level would lie at a value appropriate to the free atom, so that all the $1s$ levels would lie at the same energy. In the presence of an interaction between the atoms, this can no longer be the case. In a simple case, the coupling between nearest-neighbor atoms can be given the value A, which we now need to determine.

In general, the Schrödinger equation still provides all the information that we need to address this problem. The solution is in the approximations and boundary conditions that are applied. For the present purposes, we begin with the time-dependent form of the Schrödinger equation

$$ i\hbar \frac{\partial \psi}{\partial t} = -\frac{\hbar^2}{2m} \frac{\partial^2 \psi}{\partial x^2} + V\psi. \tag{4.31} $$

If some arbitrary function is known at a series of equally spaced points, in this case the lattice points, one can use a Taylor series to expand the function about these points and evaluate it in the regions between the known points. Consider the set of lattice points shown in Fig. 4.5 for example. Here, each point is separated from its neighbors by the value a_0 (in position). This enables us to develop a finite-difference scheme for the numerical evaluation of the Schrödinger equation. If we first expand the function in a Taylor series about the points on either side of $x_0 = ja_0$, we get

$$f(x_0 + a_0) = f(x_0) + a_0 \frac{\partial f}{\partial x}\bigg|_{x=x_0} + O(a_0^2) \rightarrow f_{j+1} \approx f_j + a_0 \frac{\partial f}{\partial x}\bigg|_j ,$$

(4.32a)

$$f(x_0 - a_0) = f(x_0) - a_0 \frac{\partial f}{\partial x}\bigg|_{x=x_0} + O(a_0^2) \rightarrow f_{j-1} \approx f_j - a_0 \frac{\partial f}{\partial x}\bigg|_j .$$

(4.32b)

The factor $O(a^2)$ is the truncation error. In the two equations on the right, we have used a shorthand notation for the *node index j*. If we now subtract the two equations on the right, we obtain an approximate form for the derivative at x_0:

$$\frac{\partial f}{\partial x}\bigg|_j = \frac{f_{j+1} - f_{j-1}}{2a_0} .$$

(4.33)

We can as easily take an average value in between, and rewrite (4.16) as

$$\frac{\partial f}{\partial x}\bigg|_{j+1/2} = \frac{f_{j+1} - f_j}{a_0} , \quad \frac{\partial f}{\partial x}\bigg|_{j-1/2} = \frac{f_j - f_{j-1}}{a_0} .$$

(4.34)

These last two forms are important for now developing the second derivative of the function, as

$$\frac{\partial^2 f}{\partial x^2}\bigg|_j = \frac{\frac{\partial f}{\partial x}\bigg|_{j+1/2} - \frac{\partial f}{\partial x}\bigg|_{j-1/2}}{a_0} = \frac{1}{a_0}\left(\frac{f_{j+1} - f_j}{a_0} - \frac{f_j - f_{j-1}}{a_0}\right)$$

(4.35)

$$= \frac{f_{j+1} + f_{j-1} - 2f_j}{a_0^2} .$$

Hence, the Schrödinger equation can now be written as

$$-\frac{\hbar^2}{2ma_0^2}(\psi_{j+1} + \psi_{j-1} - 2\psi_j) + V_j\psi_j = i\hbar\frac{\partial\psi_j}{\partial t}. \qquad (4.36)$$

We can now introduce two constants as

$$E_1 = -\frac{\hbar^2}{ma_0^2} + V \quad \text{and} \quad A = \frac{\hbar^2}{2ma_0^2}, \qquad (4.37)$$

under the assumption that all $V_j = V$. That is, the atomic potentials are all the same.

To proceed further, we now assume that the electronic states can be described by a time varying wave in which the phase is related to the total energy of the particle. Hence, we write the wave function as

$$\psi_j = K_j e^{-iEt/\hbar}. \qquad (4.38)$$

If this form is used in (4.36), we may now write the Schrödinger equation at site j as

$$EK_j = E_1 K_j - A(K_{j+1} + K_{j-1}). \qquad (4.39)$$

If we reinsert the position variation through $K_j = K(x_j)$, (4.39) can be written as

$$\begin{aligned} EK(x_j) &= E_1 K(x_j) - A\lfloor K(x_{j+1}) + K(x_{j-1})\rfloor \\ &= E_1 K(x_j) - A[K(x_j + a_0) + K(x_j - a_0)]. \end{aligned} \qquad (4.40)$$

At this point, it is fruitful to recall that our waves are plane waves, so that it can be assumed that

$$K(x) = e^{ikx}. \qquad (4.41)$$

We now find that (4.40) can be rewritten as

$$Ee^{ikx_j} = E_1 e^{ikx_j} - A\lfloor e^{ik(x_j + a_0)} + e^{ik(x_j - a_0)}\rfloor, \qquad (4.42)$$

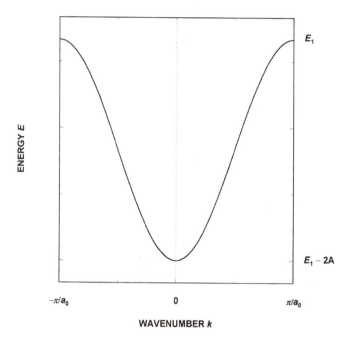

Figure 4.8. Energy variation in the lowest band obtained by plotting the form of Eq. (4.43). The energy varies sinusoidally in the band as the electron wavevector is varied.

or

$$E = E_1 - 2A \cos(ka_0). \qquad (4.43)$$

Our result is that the energy band has a cosinusoidal shape. In the case of the lowest band, the minimum is at $k = 0$, and the maxima are at $k = \pm \pi/a_0$. In the second band, the coefficient must flip sign so that the maximum is at $k = 0$, but this also requires a larger value for E_1 (which must be assumed to be E_2 for consistency). We sketch the lowest band in Fig. 4.8. In Fig. 4.9, we plot two bands, where the second band has been reflected from higher momentum states by comparison with the extended zone scheme of Fig 4.3. It is clear that the energy gap is not defined by this scheme, except as it relates to the differences between E_1 and E_2, and the band widths $2A$. Moreover, it is also clear that near the band extrema, the shape of the

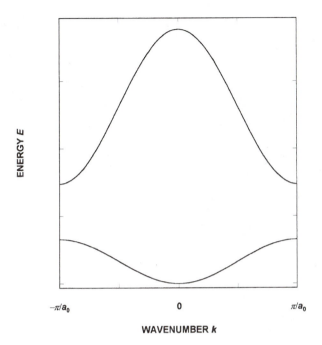

Figure 4.9. The first two energy bands obtained from the tight-binding approach are plotted here in the reduced-zone scheme.

energy band is considerably different from the free electron situation. It is in these regions that we need to define a new "effective" value for the "mass" of the electron (as a particle).

4.5. THREE DIMENSIONS AND THE BAND STRUCTURE FOR Si AND GaAs

The preceding discussion was for a single linear dimension. What happens in the real world of three dimensions? Each atom in the diamond structure, for example, is connected to four nearest neighbors, as was shown in Fig. 4.1. The bonds to these four neighbors form the tetrahedral structure of the lattice. In each of these four directions, the bond is one of the sp^3 that we already discussed. This tetrahedral directionality of the bonds automatically makes the resulting band structure three dimensional in nature. Indeed, along any single direction, such as [100], the transverse directions still play an important role. We must find how best to plot this resulting band structure. The answer lies in the *reciprocal lattice*—the values of wave vector **k**, which lie in the three-dimensional first Brillouin zone. For the face-centered-cubic lattice that is the basis of the diamond (and zinc-blende) structure, this first Brillouin zone is a truncated body-centered structure. It is truncated, since the point [000] is equivalent to the point [111] by a lattice translation of $2\pi/a_0$ in each of three directions. This is also true for any such translation along the principle axes. This is why the structure is truncated. The resulting first Brillouin zone for the diamond and zinc-blende lattices is shown in the left panel of Fig. 4.10, where we indicate the axes by k-vectors along principle axes. We recall that a direction in which **k** ∥ [100] means a direction normal to a (100) plane of the real lattice. There are some special points, which have come to have common names. The middle of the Brillouin zone, where $k = 0$, is termed the Γ point. The edge of the zone, along the [100] direction is termed the X point, and the edge of the zone along the [111] direction is termed the L point. In plotting the band structure, we usually just follow the primary symmetry axes. This leads to the most common plots in which one begins at L and follows the [111] line to Γ. One then plots along the [100] line to X. At this point, it is important to note that the point $(\pi/a_0, \pi/a_0, 0)$ is at the center of a small square area, which is actually identical to the square area that has the X point as its center. Hence, this is actually another X point, but in the second zone, and this is shown in the right-hand panel of Fig. 4.10. Thus, the common practice is to finally plot the band structure along [110] from this second X point back to Γ. The band structure of Si, following this scheme, is shown in Fig. 4.11. As is common in these materials, the maximum of the bonding band—the valence band— lies at Γ, which is the center of the zone. At this point, the bands are primarily derived from the p-states, so this point is triply degenerate (three bands converge at this point). On the other hand, the minimum of the antibonding band—the conduction band—lies along the [100], approximately 85% of the wave to X. Because there are six equivalent directions of this nature ([100], [$\bar{1}$00], [010], [0$\bar{1}$0], [001], [00$\bar{1}$]), there are six equivalent minima in the conduction band. We will discuss this in more detail in the next chapter.

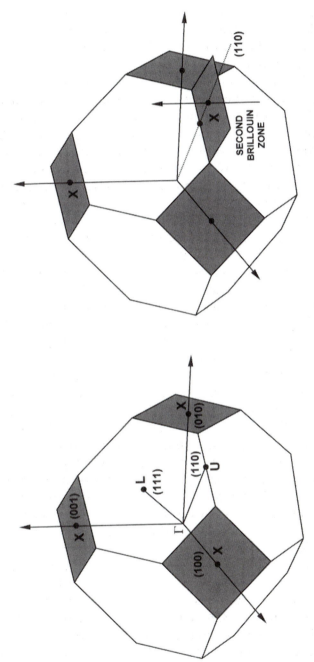

Figure 4.10. The first Brillouin zone for the diamond and zinc-blende lattices is that of the face-centered-cubic lattice. In the left figure, we define some of the important crystal directions for this lattice. Shown right is a plane that occupies the second Brillouin zone; it is placed here to indicate how the (110) axis passes from the first zone to another X point in the second zone.

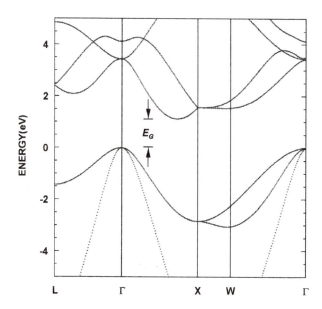

Figure 4.11. The calculated band structure of Si in the vicinity of the band gap. The *indirect* band gap is also shown for reference.

In Fig 4.12, the band structure is shown for GaAs. Here, the valence band is quite similar to that of Si, but the conduction band minimum is now at Γ as well. At this point, these conduction band states are primarily s-states. Only a single band appears in the conduction band. In fact, in nearly all tetrahedrally coordinated semiconductors, the maximum of the valence band lies at Γ. However, there are minima of the conduction band at Γ, X, and L. Which of these happens to be the actual minimum of the conduction band (or when the minima occur slightly away from X, as in Si) depends crucially on the details of the nearest-neighbor coupling. A treatment of this is clearly beyond the level of understanding we need at this point. Rather, it is important only to understand that these band structures exist, and that they can be calculated in some detail by sophisticated techniques. In Table 4.1, some of the energy gaps, relative to the top of the valence band at Γ, are listed.

We already remarked in Chapter 3 that the valence electrons in Si would be just sufficient to fill all the states of the valence band (the

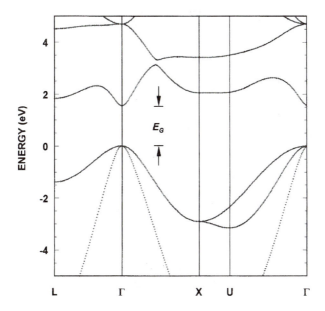

Figure 4.12. The calculated band structure of GaAs in the vicinity of the band gap. The *direct* band gap is also shown for reference.

TABLE 4.1. Some of the Principal Energy Gaps

Material	E_Γ (eV)	E_X (eV)	E_L (eV)
AlAs	2.96	1.6	2.11
GaP	2.75	2.59	2.64
GaAs	1.42	1.89	1.72
GaSb	0.7	0.89	0.86
InP	1.35	2.33	1.85
InAs	0.35	1.37	1.07
InSb	0.2	0.52	0.56
Si	4.1	1.08[a]	2.3
Ge	0.9	1.4	0.7

[a] Minimum is at 0.85 X (see Fig. 4.9).

bonding band). This is also true for GaAs, and indeed is true for any of the diamond and zinc-blende structures. The result is that all the states in the valence band are filled, while all the states in the conduction band are empty. Since the band gap is of the order of 1.1 eV for Si, this means that the material is an insulator at room temperature ($k_B T/e = 0.0259$ eV at 300 K). At higher temperatures, however, electrons can be excited from the valence band to the conduction band, leaving empty states behind. We will discuss these states in the next section, and the statistical properties of how these, and the conduction, states are filled as a function of temperature. The property of being an "insulator" at room temperature is important for the use of these materials for electronic applications, as we can modify their properties in very specific manners by the introduction of "impurity" atoms — atoms with either more than four electrons in the outer shell, or fewer than four electrons in the outer shell for Si. We discuss this in detail in Chapter 5.

4.6. EFFECTIVE MASS OF THE ELECTRON

Now let us turn to the problem of trying to establish an *effective mass* for the electrons in these complicated band structures. For this purpose, we will use the simple cosinusoidal bands pictured in Figs. 4.8 and 4.9. These bands look very little like the free electron parabolas which were shown as the dashed curves in Fig. 4.3, especially near the top of the band where the energy gap opens. Hence, the mass of an electron in such bands will not be the same as the free-electron mass. Rather, this effective mass will arise from consideration not only of external forces, but the forces of the atomic potentials that act upon the electron. Thus, we seek an effective mass that describes the response to the external force without detailing just how the extra interaction with the atomic potentials occurs. This latter is buried within the effective mass.

The basic point we need to understand is that the momentum and the wavelength are still connected by the de Broglie relationship. Now, however, the momentum is defined in terms of the band structure and the wavelength is defined by the crystal momentum, which is a result of the periodic nature of the potentials. Our

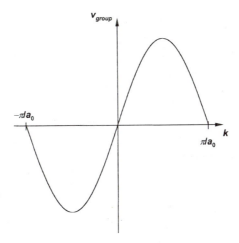

Figure 4.13. The variation of the electron velocity as a function of wave number within the first energy band, as obtained from Eq. (4.45).

beginning point is the group velocity of the particle, which still arises from the discussion of Chapter 2, as

$$v_{\text{group}} = \frac{\partial \omega}{\partial k} = \frac{1}{\hbar} \frac{\partial E}{\partial k}. \tag{4.44}$$

In the last form, we have introduced the Planck relation between frequency and energy. We have reverted to a single spatial dimension for clarity. Relationship (4.44) must still be true in our present band structure. If we use (4.43) for the energy, within the first Brillouin zone, then the group velocity is

$$v_{\text{group}} = \frac{1}{\hbar} \frac{\partial [E_1 - 2A \cos(ka_0)]}{\partial k} = \frac{2Aa_0}{\hbar} \sin(ka_0). \tag{4.45}$$

In Fig. 4.13, this velocity is plotted as a function of the crystal momentum within the first Brillouin zone. It is clear that the group velocity is a finite (noninfinite) quantity throughout the zone, and vanishes only at the zone center and at the zone edge. The de Broglie relationship assures that the momentum within the crystal is related to the wavelength through

$$p = \frac{h}{\lambda} = \frac{hk}{2\pi} = \hbar k. \tag{4.46}$$

Now, however, the wave vector k is single valued only within a single Brillouin zone. But, the group velocity in (4.45) is similarly restricted to being single valued within a single Brillouin zone. Thus, so long as we remain within a single band, and a single (the first) Brillouin zone, we may define an effective mass through (4.46) and (4.44) as

$$m^* v_{\text{group}} = \hbar k, \tag{4.47}$$

or

$$\frac{1}{m^*} = \frac{1}{\hbar^2 k} \frac{\partial E}{\partial k}. \tag{4.48}$$

In these two equations, the effective mass has been denoted as m^*, to distinguish it from the free electron mass $m_0 = 9.1 \times 10^{-31}$ kg. The definition (4.47) has several satisfying features. First, we note that the right-hand side has a zero only at $k = 0$ where the derivative of the energy also vanishes *quadratically*. As pointed out above, the group velocity is also finite, and vanishes only at $k = 0$ and $k = \pm \pi/a_0$. Since the velocity vanishes at $k = \pm \pi/a_0$, (4.48) leads to the mass diverging at these values. But this divergence is required for the left-hand side of (4.47) to have a finite value (zero times infinity is indeterminant, and must be evaluated by limiting processes, which lead to the right-hand side).

We note that one often sees a different definition of the effective mass. The beginning point for this is to start with what one thinks of as Newton's law:

$$m^* \frac{dv_{\text{group}}}{dt} = m^* \frac{\partial v_{\text{group}}}{\partial k} \frac{dk}{dt} = \hbar \frac{dk}{dt} = \text{Force}. \tag{4.49}$$

Using (4.44), this leads to

$$\frac{1}{m^*} = \frac{1}{\hbar^2} \frac{\partial^2 E}{\partial k^2}. \tag{4.50}$$

But, there is an immediate problem with this approach. The result (4.50), with (4.43) for the energy, leads to a divergence of the mass at $\pm \pi/2a_0$. But, this cannot satisfy (4.47) because both the velocity and wave vector k are finite and nonzero. We cannot use a divergent mass in this formula as it violates the de Broglie relationship! The problem lies in the initial formulation (4.49). The left-hand side is *not* Newton's equation. Newton's equation arises for the time rate of change of the *momentum*, not the velocity. To bring the mass outside the derivative is to assume that it is constant with wave vector, which is not the case at all. On the other hand, if we expand (4.43) around the minimum of the band, we can write it as

$$E \cong E_1 - 2A \left(1 - \frac{k^2 a_0^2}{2} \right) = (E_1 - 2A) + Ak^2 a_0^2. \qquad (4.51)$$

Within the limited range of validity of this expression, both (4.48) and (4.50) yield the same value of

$$\frac{1}{m^*} = \frac{2Aa_0^2}{\hbar^2} \cdot |ka_0| \ll 1. \qquad (4.52)$$

It is because of this fact that the use of (4.50) has persisted in the literature, without any caveates about its very limited range of applicability. It is an incorrect form, but its use does not lead to error provided that the parabolic band approximation (4.51) is valid.

In the previous section, we discussed how the number of electrons provided by the atoms in the diamond (or zinc-blende) lattice completely filled the states available in the valence band at low temperature. This has a significant impact on the ability of the material to conduct current. The current itself is the summation of the current carried by each individual electron. That is we sum over the current carried by the electrons, where the sum runs over the filled states in a band (in this case, the valence band). Each electron carries a current determined by its velocity and charge — the current is $-ev$ for each electron. However, the velocity varies with the energy of the electron, depending on what the value of its energy, according to (4.45). Thus, we may write the current as

$$J = -e \sum_{\substack{\text{filled} \\ \text{states}}} v_i, \qquad (4.53)$$

where the interpretation is that the index i runs over the filled states. When there are only a few filled states, this formula is easy to utilize. However, in the case of the valence band, where empty states appear at reasonably high temperatures, the formula is more difficult to evaluate. To facilitate this, we add and subtract the current of the empty states — the nonexistent current carrying particles. This is done by

$$J = -e \sum_{\substack{\text{all} \\ \text{states}}} v_i + e \sum_{\substack{\text{empty} \\ \text{states}}} v_i. \qquad (4.54)$$

A full band cannot carry current! That is, for each occupied state with momentum k, and velocity v, there is a state with $-k$, and $-v$. As a result, the first sum in (4.54) is identically zero.

What are we to make of this current in (4.54) carried by empty states? First, it can be noted that the sign of the charge (and therefore the current) has changed. Instead of a negatively charged electron, the empty state carries a positive charge. In nearly full bands, this leads us to conclude that we can more effectively compute currents by the empty states, rather than the full states, but these empty states carry a positive charge. We term these states "holes." The empty state is treated as an absence of electrons (or an absence of matter in the more normal concept of a hole). The purpose of discussing this now is to decide what mass this hole should have. We have already seen that the electron (or filled state) possesses an effective mass, and so we should expect to assign an effective mass to the hole. However, we first must decide how to describe the energy for the hole. To be sure, the empty states will sit at the top of the valence band — in Fig. 4.9, these states appear near the point $k = \pm \pi/a_0$. As the electrons gain energy, presumably from an applied field, the empty states move downward in energy. Thus, they gain energy in the downward direction, as shown in Fig. 4.14. In essence, the sign of the energy changes along with the sign of the particle. Electrons have negative

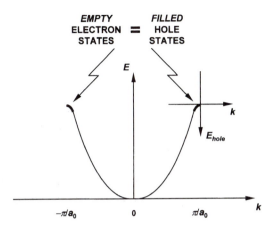

Figure 4.14. The definition of hole wave number k and energy E_{hole}. As shown, the hole states are basically empty electron states at the top of the band and increasing hole energy corresponds to moving *downward* within the band.

charge and gain energy in the upward direction of Fig. 4.9, while holes have positive charge and gain energy in the downward direction. If we expand (4.43) around the point $\pm\pi/a_0$, we can write the energy as

$$
\begin{aligned}
E &\cong E - 2A \cos((k - \pi/a_0)a_0 + \pi) \\
&= E_1 + 2A\left(1 - \frac{(k - \pi/a_0)^2 a_0^2}{2}\right) = (E_1 + 2A) - A(k - \pi/a_0)^2 a_0^2.
\end{aligned}
\tag{4.55}
$$

In this simple approximation to a parabolic band, the mass may be seen to change sign *in terms of electron energy*. However, if we interpret the hole energy as increasing with a downward motion, the mass of the hole is positive.

For a more general treatment of the mass, along the lines of the result (4.45), we may write the hole velocity as

$$
v_{\text{hole}} = -\frac{1}{\hbar}\frac{\partial E}{\partial(k - \pi/a_0)},
\tag{4.56}
$$

since the velocity for both electrons and holes vanishes at the zone edge. The minus sign appears because the energy E represents *electron* energy, and it is the negative of this quantity that represents hole energy. Using this, and the general connection of the de Broglie relation for the hole, we have that

$$\frac{1}{m^*_{\text{hole}}} = -\frac{1}{\hbar^2(k - \pi/a_0)} \frac{\partial E}{\partial(k - \pi/a_0)}. \tag{4.57}$$

The shifted momentum appears in this equation since the crystal momentum for the hole is measured from the zone boundary $k = \pm \pi/a_0$. Using (4.43), we find that the effective mass is quite like the electron mass. However, it may be compared with the result (4.55), by which we realize that the mass is for values near the band edge and tends toward infinity at the zone center, where the hole velocity must vanish. In Fig. 4.15, we plot the hole velocity, and note that it is essentially the same as Fig. 4.13. The shift in origin of the crystal momentum is balanced by the leading change of sign. In Fig. 4.16, the mass of the electrons and holes is plotted. Here, it can be seen that the parabolic mass for the electrons and holes arises at

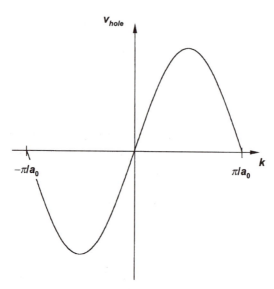

Figure 4.15. The variation of the hole velocity as a function of wave number within the first energy band, as obtained from Eq. (4.56). Note how this curve is basically identical to that plotted for electrons in Fig. 4.11.

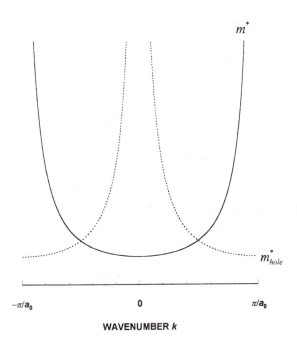

Figure 4.16. The variation of the electron and hole effective masses as a function of wave-number within the first energy band, as obtained from Eqs. (4.48) and (4.57). The electron mass diverges as the band edge is approached and electrons become diffracted by the crystal structure.

different points in the zone, according to where these two particles have their extrema.

The velocity of the particle is the same regardless of the sign of the particle! For electrons, where the minimum of the energy is at $k = 0$, positive velocity arises for displacement in the $+k$ direction. For the holes, where the "minimum" of the hole energy is at $k = \pm \pi/a_0$, positive velocity arises for displacement in the $-k$ direction. Thus, the vector velocity ($-k$ direction) for the holes is pointed in the opposite direction from the vector velocity of the electrons ($+k$ direction). Since the sign of the charge is opposite, both give a contribution to the current with the same sign. Thus positive current arises from electrons and holes moving in opposite directions, according to their individual charges. This is a proper result, since the treatment of the particles as either electron or hole is arbitrary, according to (4.53) and (4.54). Thus, we must achieve the proper current, which is a measurable quantity, regardless of which format we chose for the particles. However, care is required to properly define internal parameters, such as the crystal momentum and particle energy.

4.7. ALLOYS AND HETEROSTRUCTURES

In some sense, materials with the zinc-blende lattice are a perfect alloy of two materials, for example, Ga and As, which combine to form a new material with different properties. Both Ga and As are metals, but the material GaAs is a semiconductor with a completely different band structure from either. Similarly, AlAs is a semiconductor with a completely different band structure than either that of Al or As. What if we combine GaAs and AlAs? In general, we talk about GaAs and AlAs as being $A^{III}B^V$ compounds with one A atom and one B atom comprising the basis of the face-centered-cubic lattice. Each A atom has four B neighbors, and each B atom has four A neighbors, as can be seen in Fig. 2.16. In both GaAs and AlAs, the B atom is As. The alloy $Ga_xAl_{1-x}As$ is made by putting Ga and Al on the A atom site randomly with the "average" fraction x of Ga atoms and $1 - x$ Al atoms. In the random alloy approximation, these two materials create a system in which the properties (effective masses, band structure, lattice constant, etc.) vary smoothly from one constituent material to the other as x is varied.

The band structure of GaAs was shown in Fig. 4.12. The band structure of InAs has precisely the same direct gap form, but with different values for the various minima of the conduction band, as measured from the top of the valence band. For example, the band gap of GaAs at the Γ point is 1.4 eV, while that of InAs is 0.4 eV. In Fig. 4.17, we plot the band gap at the Γ point for the alloy $In_xGa_{1-x}As$. It can be seen that the band gap does vary smoothly from pure GaAs to pure InAs, but not in a linear manner. This deviation from linearity is called *bowing*. The band gap energy can be written as

$$E_{InGaAs} = E_{GaAs}(1 - x) + E_{InAs}x - E_B\, x(1 - x)$$

$$= 0.4 + 1.0(1 - x) - (0.37 \pm 0.09)x(1 - x). \tag{4.58}$$

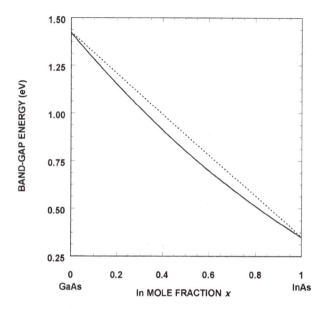

Figure 4.17. The variation of the energy-band gap as a function of the indium mole fraction in InGaAs. The dotted line is the result obtained from the virtual-crystal approximation, while the solid line is the experimental curve and clearly "bows" away from the predictions of theory.

The value of the bowing parameter is an experimentally determined quantity, as no theory currently exists that can predict this quantity satisfactorily. The source of the bowing is partly due to the nonlinearity in the hybrid energies that give rise to the band gap and partly due to deviations from the random alloy theory. While the lattice constant varies linearly between that of GaAs and InAs, the spacing between the atoms on the basis deviates from this linear behavior. That is, the In and As atoms prefer to maintain their InAs spacing instead of changing as the alloy requires. This causes a random potential distortion in the lattice, which is thought to contribute to the bowing of the energy gap. InGaAs is very useful for infrared applications, since choosing a composition $x \sim 0.47$ gives a band gap at ~ 0.8 eV, which is the correct energy to use for optical fiber communications. Table 4.2 lists some of the bowing energies for various alloys.

While InAs and GaAs both were direct gap materials (the minimum of the conduction band and the maximum of the valence

TABLE 4.2. Energy Gap Bowling Parameters

System	E_B (eV)
GaP–GaAs	0.19
GaP–InP	0.65
GaP–GaSb	1.3
GaAs–InAs	0.37
GaAs–AlAs	0.4
GaAs–GaSb	0.8
GaSb–InSb	0.5
InP–InAs	0.1
InP–InSb	1.6
InAs–AlAs	0.4

band occurred at the same point in the Brillouin zone), this is not the case for GaAs and AlAs. AlAs is an indirect gap material, in which the minimum of the conduction band is at the X point ([100]). In AlAs, the direct gap at Γ has a value of 2.4 eV, while the separation of the X minimum and the Γ maximum is only 2.2 eV. Thus, the band gap in the alloy $Al_xGa_{1-x}As$ does not vary smoothly, but shows a "kink" at about $x = 0.4$, where the material changes from a direct to an indirect band gap. This is shown in Fig. 4.18, where we plot the energies of the various band gaps for the alloy. Visible light has energies above approximately 1.75 eV, and this value can be obtained in an alloy with $\sim 25\%$ AlAs. This can be used for optical applications in the red end of the spectrum (laser diodes, for example).

The alloying of two zinc-blende materials to make a new material with different, but desired, properties can be extended to alloying on both sublattices. One such material is $In_xGa_{1-x}As_yP_{1-y}$. Such a material is called a *quaternary* compound, while InGaAs is a *ternary* compound. The two compositions x and y can be varied to match not only a desired band gap, but also to achieve a desired lattice constant. The latter is needed in order to grow the material on a particular substrate without introducing strain and defects. This variation of alloy compositions, in order to achieve a set of desired crystal properties, is called *band-gap engineering*. With modern crys-

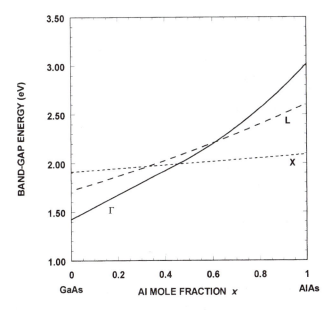

Figure 4.18. The variation of the different energy-band gaps as a function of the aluminum mole fraction in AlGaAs. Above a molar fraction $x \approx 0.45$ the alloy crosses over from a direct to an indirect semiconductor.

tal growth techniques, band-gap engineering of the zinc-blende materials is heavily used for optical and microwave applications. Even Si has become a base for such band-gap engineering, as the alloy $Si_x Ge_{1-x}$ has become interesting for microwave applications. The ability to grow these materials on an arbitrary, but lattice-matched, substrate has further led to *heterostructures* in which different materials are layered on one another.

By heterostructure, we specifically mean two different materials which are adjoined to make a single structure. The example most commonly found is that of GaAs and AlGaAs. The lattice constant of AlAs is very nearly the same as that of GaAs, so that the alloy has this same lattice constant across the entire range of x. Thus, it is quite simple to grow a layer of AlGaAs on top of a layer of GaAs, especially by molecular-beam epitaxy, a method in which crystals are grown one atomic layer at a time. Constructing such materials is especially important in band-gap engineering of special electronic

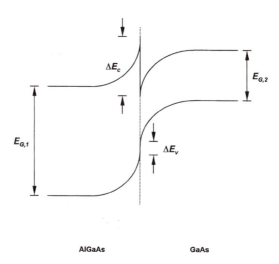

Figure 4.19. Variation of the semiconductor energy bands across the interface of a GaAs–AlGaAs heterostructure. The magnitude of the band bending at the interface depends on the doping level of the two materials and here we have assumed that the AlGaAs (left) is more strongly n-type doped than the GaAs. Also indicated on the figure are the conduction- and valence-band offsets at the interface and the bulk energy gaps.

structures. In Fig. 4.19, we show the conduction band minima and the valence band maxima for two direct band gap semiconductors which are joined to form a heterostructure. The band gap of material 1 is different than that of material 2. However, the Fermi energy must be constant through the structure (we will see this in detail in later chapters) if no current is to flow. Hence, the bands must bend at the interface between the materials to accommodate the *band offsets*. The difference between the two band gaps is accommodated by one part appearing as a difference between the two conduction band minima. This is termed the conduction band offset ΔE_c. The remainder of the band gap difference is in the valence band, and this is termed the valence band offset ΔE_v. Obviously,

$$\Delta E_c + \Delta E_v = |E_{G,1} - E_{G,2}| = \Delta E_G. \tag{4.59}$$

However, the determination of the individual band offsets is one of experiment—there exists no theory sufficiently accurate to predict this offset (we should note that there are several theories which *claim* to be able to do this).

The offsets for the GaAs–AlGaAs interface are the best studied, and it is thought that 67% of ΔE_G appears as ΔE_c. The band gap at the Γ point for the alloy $Al_x Ga_{1-x}As$ is thought to be given by

$$E_\Gamma(x) = 1.4 + 0.6(1 - x) - 0.4x(1 - x), \qquad (4.60)$$

so that for $x = 0.3$, the band gap is a direct one with energy of 1.74 eV. This means that the conduction band offset is about 225 meV, while the valence band offset is 110 meV. Even if both materials are intrinsic, so that the Fermi level is at the center of the gap, there is band bending due to this disparity in the two band offsets. However, if the two materials are configured to have the same Fermi energy (this is explained in the next chapter), then we can arrange for there to be no band bending in the conduction band. Hence, for a layer of AlGaAs placed between two GaAs layers, the conduction band energy appears as in Fig. 4.20. A barrier exists between electrons on

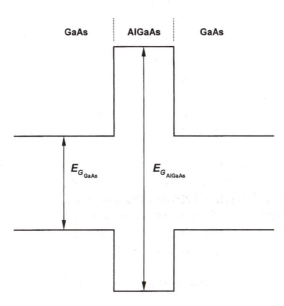

Figure 4.20. By sandwiching AlGaAs between two layers of GaAs, an energy barrier to carrier motion can be formed. The barrier arises due to the difference in band gap between the two materials. Note in this figure how the band gap discontinuity induces a barrier to electron motion in the conduction band and to hole motion in the valence band. Such energy barriers are frequently employed in actual device structures, such as resonant-tunneling diodes.

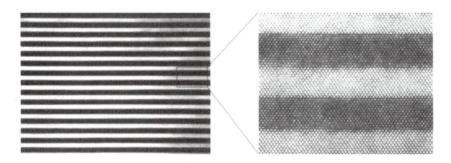

Figure 4.21. Scanning electron micrograph of a multilayer GaAs–AlGaAs superlattice. Shown left is a sequence of nearly thirty different layers, while on the right-hand side the individual atomic resolution is indicated. *Picture provided courtesy of A. Trampert and O. Brandt, Paul-Drude Institute, Berlin.*

one side and electrons on the other side. In fact, this barrier has precisely the form as that in Fig. 3.12. Thus, with modern layer-by-layer growth techniques and band-gap engineering of alloys and heterostructures, we can fabricate any of the tunneling barriers discussed in the last chapter, and can prepare special structures for unique applications, some of which will be discussed in Chapter 6. Indeed, by alternating layers of the alloy AlGaAs with layers of GaAs, one can construct a *superlattice* with potential shape as shown in Fig. 4.6. Instead of atomic potentials, now the potentials are created in the underlying materials by using discontinuities in the conduction (and valence) bands. In this way, artificial materials may be created with interesting new properties. In Fig. 4.21, a high resolution electron micrograph of laternating layers of GaAs and AlAs is shown. Each layer is only a few nanometers thick, and the structure is used to study the superlattice band structure by optical techniques.

4.8. THE ATOMS IN MOTION

The concept of sound waves in solids has been around almost as long as solids have been studied. Although it was assumed in the previous sections that the lattice was a rigid lattice, this in fact is not the case. Rather, the atoms themselves have a thermal motion, which leads to *vibrations* of the lattice. These vibrations themselves can act to scatter the electrons away from their simple plane wave states. That is, an electron wave can be moving through the lattice, but "strike" one of the moving atoms. Energy and momentum can be transferred through this process.

The motion of the various atoms is much like the motion of the electrons, with the important exception that the atoms are forced, on the average, to retain their lattice positions. The lattice, of course, is a three-dimensional system. However, when the lattice motion is a wave along one of the principal axes of the lattice, an approximation of a one-dimensional chain is quite important. Although this is a simple model, its applicability can be extended to the real crystal if each atom represents the typical motion of an entire atomic plane perpendicular to the wave motion.

Consider first a one-dimensional chain of atoms that constitutes a one-dimensional lattice. At rest, the atoms are separated a distance a_0, as shown in Fig. 4.22. Each atom is assumed to have a mass M, and all of the atoms are assumed to be identical (otherwise, it would not be the simple lattice shown in this figure). Our goal here is to show that the solutions are waves, and to find the *dispersion relations* for these waves. In Fig. 4.22, the displacement of the central atom has been shown. In reality, each of the atoms is displaced along the chain, and it will be assumed that the sth atom is displaced by an amount u_s. The interatomic forces work to restore this displacement to the equilibrium position, just as if there were springs attached between each of the atoms.

In the approach here, only the forces between each nearest-neighbor atom will be considered to be significant, and these will

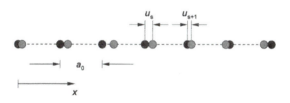

Figure 4.22. Displacements of atoms in a one-dimensional crystal. The black symbols denote the equilibrium positions of atoms in the crystals, while the gray symbols indicate the different atomic displacements at a given point in time. The symbol u_s is used to denote the displacement of the sth atom from its equilibrium position in the crystal.

actually be taken in the simple quadratic limit (which is really the linear spring limit). Then, one can immediately write down the differential equation for the motion of the sth atom as

$$M \frac{d^2 u_s}{dt^2} = F_s = C(u_{s+1} - u_s) - C(u_s - u_{s-1}),$$ (4.61)

where C is the force constant for the interatomic springs. Since the current interest is in lattice wave motion, it will be assumed that a sinusoidal steady state exists, in which the time variation is given by $\exp(-i\omega t)$, so that the left-hand side of (4.61) becomes $-M\omega^2 u_s$. In addition, we take the wave motion to have the positional form $\exp(iqx)$, where q is the wave (momentum) vector for the lattice wave. Here, q is fully analogous to k used above. With this approximation, we find that

$$u_{s\pm 1} \sim u_s e^{\pm iqa_0}.$$ (4.62)

Using this in (4.61), we may write the solutions as

$$-M\omega^2 u_s = Cu_s(e^{iqa_0} + e^{-iqa_0} - 2),$$ (4.63)

or

$$\omega^2 = \frac{2C}{M}(1 - \cos(qa_0)).$$ (4.64)

This solution is quite like (4.43) in spirit, if not in detail. In fact, we can take the root of this last expression to write

$$\omega = 2\sqrt{\frac{C}{M}} \left| \sin\left(\frac{qa_0}{2}\right) \right|,$$ (4.65)

since we require a positive frequency. This solution is plotted in Fig. 4.23. We note immediately that there is a principal range of q, given

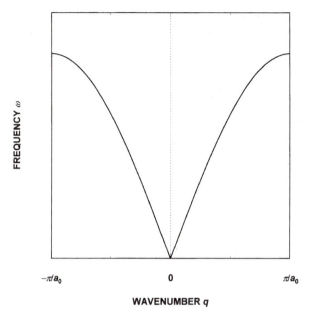

Figure 4.23. Dispersion curve for lattice waves in a one-dimensional crystal.

by

$$-\frac{\pi}{a_0} < q < \frac{\pi}{a_0}, \tag{4.66}$$

just as for the electrons. This region is recognized as the first Brillouin zone. Again, the lattice defines the range of the wave vector (here it is q instead of k, but the result is the same). While we have not talked about the boundaries of the linear chain, it is clear from the result that periodic boundary conditions are an integral part of the wave solution.

For small values of the wave vector q, the sinusoid may be expanded so that the frequency is related to the wave vector through

$$\omega = \sqrt{\frac{C}{M}} q a_0. \tag{4.67}$$

This is a familiar form for elastic waves, in that the frequency is a linear function of the wave vector q. This wave now has the property of a sound wave moving through the lattice, with a characteristic velocity v_s (for sound velocity) given by

$$v_s = \frac{\partial \omega}{\partial q} = a_0 \sqrt{\frac{C}{M}}. \tag{4.68}$$

In fact, by measuring the sound velocity of such an acoustic wave through the lattice, (4.68) may be used to determine information about the spring constant C. This wave is called an acoustic wave, since the frequency is quite low and the wavelength is quite long.

It is important to note, however, that Si, and the various zinc-blende compounds, have *two* atoms per lattice site — these two atoms form the basis of the lattice. This means that the linear chain of atoms, studied above, should have two atoms per lattice spacing a_0. In the case of the zinc-blende materials, these two atoms are dissimilar — one may be Ga and the other As, for example. Thus, we now turn our attention to the slightly more difficult problem of a diatomic linear chain in which there is a basis of two atoms per lattice site. This is a linear chain in which the two atoms alternate, either in spacing or in mass. We will treat the case in which the masses are different, but the case for which the spacings are different (different spring constants) is worked out as easily. We consider Fig. 4.24, which depicts this diatomic linear chain. Even values of the site index s refer to one type of atom, with mass M_1. Odd values of the site index s refer to the other atom with mass M_2. For the even atoms, we still refer to the displacement with the variable u_s. However, for

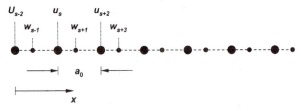

Figure 4.24. A schematic illustration of a crystal with a basis. In this case, the one-dimensional crystal structure consists of a cubic lattice, to each point of which a diatomic basis of atoms is attached. In the figure shown here, the basis is sketched as consisting of two different atomic species.

the odd atoms, we refer to the displacement with the variable w_s. The lattice constant a_0 is now the distance between *like* atoms. This is different from the simple distance between the atoms. Nevertheless, we may write down the equations for the odd and even atoms, in analogy with (4.61), as

$$M_1 \frac{d^2 u_s}{dt^2} = C(w_{s+1} - u_s) - C(u_s - w_{s-1})$$

$$M_2 \frac{d^2 w_{s-1}}{dt^2} = C(u_s - w_{s-1}) - C(u_{s-2} - w_{s-1}). \qquad (4.69)$$

As in the previous case, we take the lattice displacements as waves in which

$$u_s, w_{s-1} \sim e^{i(qx - \omega t)}. \qquad (4.70)$$

Now, equations (4.69) become

$$-M_1 \omega^2 u_s = C(1 + e^{iqa_0})w_{s-1} - 2C u_s,$$

$$-M_2 \omega^2 w_{s-1} = C(1 + e^{-iqa_0})u_s - 2C w_{s-1}. \qquad (4.71)$$

In this last equation, the displacement vectors are in term of the lattice constant because of the two displacements that are being treated. The dispersion relation is found by diagonalizing the determinant of the equations above, since there is no forcing function (therefore the determinant of the matrix must vanish, as in the Kronig–Penney model). Thus, we require

$$\begin{vmatrix} 2C - M_1 \omega^2 & -C(1 + e^{iqa_0}) \\ -C(1 + e^{-iqa_0}) & 2C - M_2 \omega^2 \end{vmatrix} = 0. \qquad 4.72)$$

This determinant yields the dispersion relation

$$\omega^4 - 2C \left(\frac{M_1 + M_2}{M_1 M_2} \right) \omega^2 + \frac{2C^2}{M_1 M_2} [1 - \cos(qa_0)] = 0. \qquad (4.73)$$

The last term retains the sinusoidal variation with crystal momentum

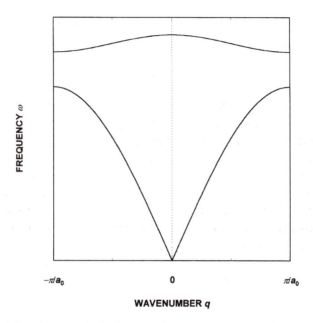

Figure 4.25. Dispersion curve for lattice waves in a one-dimensional crystal with a diatomic crystal. Note how this curve has *two* branches, which are referred to as the acoustic and optical branches.

that appeared in the case of the single atomic linear chain. However, there are two branches (solutions) for each value of ω. One is a low-frequency branch, which corresponds to the acoustic branch discussed above. The other branch, however, has very high frequency components that lie in the infrared part of the spectrum. For this reason, it is called the optical branch. We plot the two branches in Fig. 4.25.

Let us now look at the limiting frequencies of the vibrations. First, take the case of $q = 0$, which lies at the center of the Brillouin zone. The frequencies for this condition are given by the solution to

$$\omega^2 \left[\omega^2 - 2C \left(\frac{M_1 + M_2}{M_1 M_2} \right) \right] = 0. \tag{4.74}$$

Obviously, one solution has zero frequency, and this is the acoustic branch. The other mode, however, has the frequency

$$\omega = \left[2C \frac{M_1 + M_2}{M_1 M_2} \right]^{1/2}. \tag{4.75}$$

This frequency is a vibration with a reduced mass (which is the average mass of the two atoms) and represents the coupling of the two individual atoms together. This mode of vibration corresponds to a static displacement of of the atoms of mass M_2 relative to the atoms of mass M_1. That is, the two atoms of the basis are oscillating relative to one another. This condition remains for finite, but small q—the two-atom sublattices of the basis oscillate against one another. However, both atoms are in motion and that is why the reduced mass appears in the equation.

Now consider the case for which $qa_0 = \pi$. This is the zone edge. From (4.73), we can write the dispersion relation as

$$\omega^4 - 2C \left(\frac{M_1 + M_2}{M_1 M_2} \right) \omega^2 + \frac{4C^2}{M_1 M_2} = 0. \tag{4.76}$$

Again, there are two frequencies that result from this equation. These are given by

$$\omega_1 = \sqrt{\frac{2C}{M_1}}, \qquad \omega_2 = \sqrt{\frac{2C}{M_2}}. \tag{4.77}$$

These two frequencies each involve only a single atom of the pair. This is striking! The first frequency corresponds to the vibration of the chain of atoms of type M_1, and the other chain is at rest. The second frequency reverses this, the vibration is of the chain of atoms of type M_2, and the other chain is at rest. Which frequency is higher? This of course depends on the two masses. The smaller mass generates the higher frequency. This may be seen in Fig. 4.25. In the case of Si, the two masses are the same and the frequency at $qa_0 = \pi$ are the same.

If an electron interacts with an acoustic mode, the energy exchange is quite small, since the frequency of the acoustic modes is small compared to the equivalent frequency of the electrons. This scattering is mainly elastic—energy is conserved, but momentum is changed. On the other hand, if the electron interacts with the optical modes, a considerable energy is lost (or gained) by the electron — this energy goes into (or derives from) the lattice vibrations. The preceding treatment has assumed that the lattice vibrations are classical oscilllations corresponding to weights and springs. In fact, however, these oscilations should be quantized, so that the energy in a given mode is quantized according to its frequency and the Planck relation $E = \hbar\omega$. Thus, energy units corresponding to this frequency are gained and lost by the electrons in the scattering process. These units of lattice energy are termed *phonons*.[6] Since these are not Fermions — they have zero spin — they do not satisfy the Pauli exclusion principle, so there is no limit as to how many may exist in any quantum state. With interactions between the electrons and the lattice, the electron may gain or lose one phonon of energy, with the corresponding momentum change (which is determined by the appropriate branch of the lattice vibrations).

4.9. TYPES OF MATERIALS

Among the crystalline solids, we have discussed materials as metals, insulators, and semiconductors. This is the case whether we talked about amorphous materials, polycrystalline materials, or single crystal materials. How are we to distinguish between these? The answer has already appeared in the discussion. Consider Fig. 4.26. A metal has only enough outer shell electrons to partially fill the highest occupied energy band. In metals with only a single outer shell electron, this highest occupied band has $2N$ states, but there are only N electrons. Thus, the band is only one-half filled, as shown in Fig. 4.26(a). In this situation, it is very easy for the electrons in the highest occupied states to gain energy and thereby move in response to an applied electric field. Because there are so many electrons in this band, the conductivity is very high. This is the primary property of a metal, when we talk about bands and conductivity. The band is one-half filled, and the conductivity is very high due to the large number of electrons in this band.

On the other hand, in materials such as silicon (and the zincblende materials), there are just sufficient electrons to completely fill

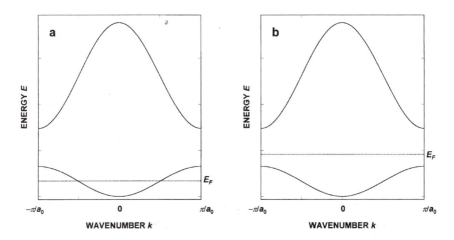

Figure 4.26. (a) The Fermi energy lies in the middle of the band for a metal, which leaves a large number of available states above this energy. (b) For an insulator, or a semiconductor, the Fermi level lies near the middle of the gap between two bands.

the highest occupied band, which is termed the valence band, as shown in Fig. 4.26(b). The next higher band (which is called the conduction band) is thus completely empty at $T = 0$. Thus, electrons in this full band cannot gain any energy from the field, as there are no states to which they easily can move, since there is a band gap to the next band. These materials are generally insulators. However, we normally discuss insulators when the band gap is large (>3 eV) so that there is no thermal excitation across the band gap. In the case of Si, the gap is only 1.08 eV so that there is a reasonable excitation of carriers across the band gap at temperatures of a few hundred Kelvin (that is, just above room temperature). For this case, there are electrons in the next highest band, and holes remain in this normally highest occupied band — the valence band. Materials which have insulating properties (full valence band, empty conduction band), but for which carriers may be thermally excited across the band gap, are called *semiconductors*. The entire world of microelectronics is based on the properties of semiconductors, and the relative ease with which these properties can be modified by the inclusion of special impurity atoms. The differences between semiconductors and insulators is one of degree, and this degree will be explored in the next chapter. There, we will discuss the statistics of the available states, and develop the probability of occupancy for any given state — the Fermi–Dirac distribution function. This latter is characterized by an important energy, the Fermi energy, which we have already encountered in the last section.

There remains one further class of material — the semimetals. These are materials in which the number of electrons would normally make the material an insulator, but the conduction band often overlaps the valence band. For example, the minimum of the conduction band may be at the zone edge, while the maximum of the valence band is at the zone center. However, the gap is negative! That is, the minimum of the conduction band lies below the maximum of the valence band, so that both holes and electrons are present. We will not deal with these materials further.

REFERENCES

1. J. M. Ziman, *Principles of the Theory of Solids* (Cambridge: Cambridge Univ. Press), Sec. 4.2.
2. W. L. Bragg, Proc. Cambridge Phil. Soc. **17**, 43 (1913).
3. L. Solymar and D. Walsh, *Electrical Properties of Materials* (New York: Oxford University Press), Ch. 5.
4. L. Brillouin, *Wave Propagation in Periodic Structures* (New York: Dover, 1953).
5. N. F. Mott and H. Jones, *Theory of the Properties of Metals and Alloys* (New York: Dover, 1958).
6. C. Kittel, *Introduction to Solid State Physics*, 6th ed. (New York: John Wiley, 1986).

PROBLEMS

1. Give a quantitative argument for why the electrons are in the center of potential wells in the periodic potential.
2. Expand the determinant that arises from (4.24)–(4.27), and show that (4.30) can be obtained from it.

3. Consider a square lattice in two dimensions, for which the lattice constant is a. Draw the first four Brillouin zones.

4. Consider the case for which the potential in the Schrödinger equation is described by

$$V(x) = V_0 + V_1 \cos(2\pi x/d).$$

The solutions to the Schrödinger equation in this case are known as Mathieu functions. These can be found by approximation. Use only V_0 and find the lowest order solution Ψ_0. Using this solution in the term with V_1 as a forcing function, find the next approximation Ψ_1. Discuss the connections between the Mathieu functions and Bloch functions.

5. In the Kronig–Penney model, show that the tops of each energy band corresponds to the energy of the corresponding level in a potential well with infinite walls.

6. An electron in a particular semiconductor is found to have energy (measured from the conduction band minimum) of

$$E = \frac{E_G}{2}\left[\sqrt{1 + \frac{2\hbar^2 k^2}{M E_G}} - 1\right].$$

Find the effective mass of the electrons, as a function of the energy. Show that at small values of energy, the square root can be expanded, and the bands become parabolic.

7. What composition of the ternary alloy InGaAs is required to have a band gap that corresponds to a photon wavelength of 1.53 μm?

8. In three dimensions, the sound velocity is more usually represented in terms of the mass density ρ_m (kg/m³). Extending (4.68) to the continuum leads to the sound velocity as $v_s = \sqrt{C/2\rho_m}$, where C is an average *stiffness constant*. If the average stiffness constant in Si, for longitudinal waves, has the value 3.89×10^{11} N/m², what is the sound velocity? Using the size of the unit cell in Si, estimate the optical phonon frequency with (4.75) (use the mass density to find the effective mass M of the two atoms in the unit cell).

CHAPTER 5

Electrical Transport

In the preceding few chapters, we have primarily discussed the crystal properties and the energy bands of the semiconductors. We pointed out that, at low temperatures, the valence band was completely full, and the conduction band was completely empty. Moreover, the full valence band could not conduct current, since there were no empty states in which to excite carriers to gain momentum from any applied electric field. In addition, the empty conduction band could not conduct current, since there were no electrons with which to carry the current. Thus, at low temperatures, the semiconductor is effectively an insulator. But, what about at high temperatures? The energy gap in most semiconductors is only of the order of 1 eV or so. Thus, at sufficiently high temperatures, thermal excitation of a few of the electrons from the valence band to the conduction band is possible. This would leave a few holes in the valence band and provide an equal number of electrons in the conduction band. Thus, both bands could conduct a current. How are we to determine just how many such electron and hole pairs are produced? Moreover, how do we determine the amount of current that is subsequently carried by these carriers? Finally, can we do something to the semiconductor to produce more electrons and/or holes? In this chapter, we will answer these questions.

To answer the first question—how many electron and hole pairs are produced at a given temperature—we must understand the occupation probability. That is, we need to seek a function $f(x, E, T)$, which gives the probability that an electron exists with position between x and $x + \Delta x$ and wave vector between k and $k + \Delta k$ at a given temperature T. The energy enters by its relationship to the wave vector $E = E(k)$; that is, the probability that an electron is in this region can be expressed as

$$P(x, k, T) = f[x, E(k), T]\Delta x \Delta k. \tag{5.1}$$

This has the comfortable result that the local electron density may be expressed as a summation over all the momentum states, as

$$n(x, T) \sim \int d^3 k f[x, E(k), T]. \tag{5.2}$$

If the wave vector was a continuous function, Eq. (5.2) would be an equality. However, since k is a discrete quantity because of the lattice, we will have to introduce a parameter/constant into the right-hand side to make this equation an equality.

Here, it is obvious that the first task is to identify the probability function f, and then to see how to evaluate the integral to compute the density of electrons (and of holes in the valence band). In the next section, the first of these two tasks is accomplished where we determine the Fermi–Dirac distribution function. We then perform the second task in the following section on intrinsic semiconductors. Following that, we will address the question about modifying the behavior of the semiconductors by the introduction of specific impurities to make extrinsic semiconductors. Once these tasks are accomplished, we will be prepared to study the transport properties in the subsequent sections.

5.1. FERMI–DIRAC STATISTICS

The one fact with which we have become familiar is that semiconductors are quantum solids. We must deal with the quantum mechanics of their properties. Hence, the desired probability function must

obey these quantum principles, notably the Pauli exclusion principle — only two electrons, with opposite spins, can occupy any single
quantum state. The desired probability function which satisfies the
Pauli principle is the Fermi–Dirac distribution[1]

$$f(E, T) = \frac{1}{1 + \exp\left(\dfrac{E - E_F}{k_B T}\right)}. \tag{5.3}$$

This probability describes the likelihood that a state of energy E is
occupied at a given temperature. This probability, in turn, depends
upon the relationship of the energy E to a new, characteristic energy,
E_F, which is called the *Fermi energy*. If $E \ll E_F$, then the probability
of the state being occupied is nearly unity. On the other hand, if
$E \gg E_F$, then the probability of the state being occupied is nearly
zero. Between these two extremes, there is a transition whose properties are sensitive to the temperature. In Fig. 5.1, we plot this
function for three different temperatures along with a conceptual
view of the conduction and valence bands. At absolute zero, the
transition is very sharp, with all available states below the Fermi
energy full and all available states above the Fermi energy empty. As
the temperature is raised, the transition becomes broad, and at a
sufficiently high temperature, some available states in the valence
band become empty while some available states in the conduction
band are filled. It is clear that this function provides the proper
behavior for semiconductors.

In arriving at (5.3), we must subject the problem to two constraints. First, the number of electrons is given by the atomic lattice
and the atoms and bonding of the solid. Hence, the total number of
electrons which fill the states must be a constant with

$$N = \sum_i n_i, \tag{5.4}$$

where the sum runs over the available states, which are indexed by
the subscript i. In addition, the total energy of the electrons must be
a constant, which we express as

$$E = \sum_i n_i E_i. \tag{5.5}$$

These two constraints will be applied in arriving at (5.3). The

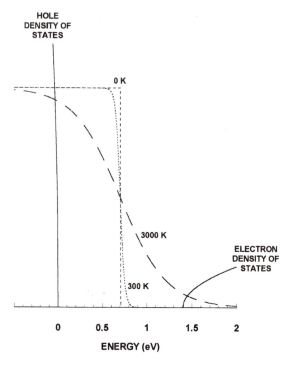

Figure 5.1. The energy-dependence of the Fermi–Dirac function is plotted for various temperatures. The density of states in the conduction and valence bands are shown as bold lines and
the differences in these two curves result from differences in the electron and hole effective mass.
At 0 K, the value of this function is one for all energies below the Fermi energy and zero for
all energies above this.

beginning place is the fact that in state i, which can have a large
degeneracy (we worry later about the Pauli principle), the number of
ways w_i in which n_i particles can be put into S_i available states in
level I is given by the binomial coefficient

$$w_i = \frac{S_i!}{n_i!(S_i - n_i)!}.\tag{5.6}$$

We can see this in the following way. The probability of putting the
first particle in box j is given by

$$P_{j1} = \frac{n_i}{S_i}.\tag{5.7}$$

The probability of the second ball also going into this box is reduced by there being one less electron and one less possible state, or

$$P_{j2} = \frac{n_i - 1}{S_i - 1}.$$ (5.8)

We can continue in this way through all n_i particles. Then the total probability of putting these particles into the box is given by

$$P_j = P_{j1}P_{j2} \cdots P_{jn_i} = \frac{n_i}{S_i} \frac{n_i - 1}{S_i - 1} \cdots \frac{1}{S_i - n_i + 1}$$

$$= \frac{n_i!(S_i - n_i)!}{S_i!} \equiv \frac{1}{w_i}$$ (5.9)

Now, the number of ways in which we can arrange all of the electrons into the available states is given by the product of the w_i over the available number of electrons. This can be written as

$$W = w_1 w_2 \cdots w_n = \frac{S_1!}{n_1!(S_1 - n_1)!} \frac{S_2!}{n_2!(S_2 - n_2)!} \cdots \frac{S_n!}{n_n!(S_n - n_n)!}.$$ (5.10)

This is a mess, to be sure. However, we can simplify it by taking the natural logarithm of the product. Then, the product becomes a sum of logs. Hence, we can rewrite (5.10) as

$$\ln W = \sum_i \ln \left(\frac{S_i!}{n_i!(S_i - n_i)!} \right).$$ (5.11)

To proceed further, we employ an approximation—the Stirling approximation—for the logarithm of a factorial:

$$\ln(x!) = \ln(x) + \ln(x - 1) + \cdots + \ln(1) \approx \left(x + \frac{1}{2} \right) \ln(x) - x.$$ (5.12)

Using this, we can rewrite (5.11) as

$$\ln W \approx \sum_i \left\{ \left(S_i + \frac{1}{2} \right) \ln(S_i) - \left(n_i + \frac{1}{2} \right) \ln(n_i) - \left(S_i - n_i + \frac{1}{2} \right) \ln(S_i - n_i) \right\}.$$

$$(5.13)$$

We want to minimize (with respect to n_i) the number of ways in which the distribution can occur, so we will deal with the derivative of (5.13). For example,

$$d \ln n_i = d[(n_i + 1/2) \ln n_i - n_i]$$

$$= dn_i \ln n_i + \frac{n_i + 1/2}{n_i} dn_i - dn_i \qquad (5.14)$$

$$\approx dn_i \ln n_i.$$

With this condition, we can now write (5.13) as

$$d \ln W = \sum_i [-\ln n_i + \ln(S_i - n_i)] \, dn_i = \sum_i \ln \left(\frac{S_i - n_i}{n_i} \right) dn_i. \quad (5.15)$$

We now want to employ our two constraints, which we expressed in (5.4) and (5.5). To do this, we take the differentials of these as

$$dN = 0 = \sum_i dn_i, \qquad dE = 0 = \sum_i E_i \, dn_i. \qquad (5.16)$$

We are not free to combine these in just any old manner. However, the differentials are zero, so that they can be added to (5.15) without changing the result, and this is done by multiplying each of these terms by a constant (to be determined later) and then using them in (5.13). Hence, the imposition of these constraints leads to

$$d \ln W = \sum_i \left[\ln \left(\frac{S_i - n_i}{n_i} \right) - \alpha - \beta E_i \right] dn_i. \qquad (5.17)$$

A sufficient condition that the left-hand side be zero (for the minimization) is that the term within the square brackets be zero. This condition leads to

$$\ln\left(\frac{S_i - n_i}{n_i}\right) - \alpha - \beta E_i = 0. \tag{5.18}$$

We may now solve for the probability of a state being occupied as

$$P_i = \frac{n_i}{S_i} = \frac{1}{1 + \exp(\alpha + \beta E_i)}. \tag{5.19}$$

We physically define the Fermi energy to be that energy at which the probability is exactly 1/2 at all temperatures. This leads to $\alpha = -\beta E_F$. At very high temperatures, we need to recover classical statistics, as expressed by the Maxwell–Boltzmann distribution

$$f_{\text{MB}} = e^{-E/k_B T}. \tag{5.20}$$

This leads us to recognize that $\beta = 1/k_B T$. At this point, we have recovered (5.3) as

$$f(E, T) = \frac{1}{1 + \exp\left(\dfrac{E - E_F}{k_B T}\right)}$$

The Fermi–Dirac distribution, as (5.20) is known, is a quantum mechanical distribution that gives us the probability that an electron occupies a state of energy E at any temperature T. The values that this distribution takes are between 0 and 1. Moreover, the probability that a state is empty is $1 - f(E)$. In a semiconductor, the Fermi energy at low temperature is midway between the conduction and valence bands. For Si, the band gap is 1.1 eV. So, the probability that an electron is in the conduction band at 300 K is approximately ($E - E_F \sim 0.55$ eV, $k_B T \sim 25.9$ meV) 6×10^{-10}. However, if we raise the temperature to 600 K, the probability ($k_B T \sim 51.8$ meV) becomes 2.4×10^{-5}. Thus, raising the temperature by only a factor of two raises the probability by almost 5 orders of magnitude! This is because the probability of exciting an electron from the valence band into the conduction band grows exponentially with temperature, as we can see from the Fermi–Dirac distribution.

5.2. INTRINSIC SEMICONDUCTORS

We now turn our attention to the determination of just how many electrons are in the conduction band, and how many holes are in the valence band, at any given temperature. We have estimated this with the approximation that appeared as (5.3). The proper way to do this, since the wave vector is a discrete variable is to us a summation, rather than an integral. Hence, we should write (5.3) as[2]

$$n(x, t) = \sum_k f[x, E(k), t], \tag{5.21}$$

and k runs over all three directions. In fact, it would be much easier to integrate (5.3) than to perform the summation in the above equation. We can convert the summation into an integration with the proper correction factor [which was the missing term in (5.3)]. The integral sums over an infinite number of values of k, regardless of how large the range of integration is set to be. This is because the integral assumes that k is a continuous variable. But in Chapter 4 we found that k only took discrete values, which were given by, for example,

$$k_x \sim \frac{2r\pi}{Na}, \qquad r = 0, 1, 2, \ldots, N - 1. \tag{5.22}$$

Here, N is the number of atoms in the one-dimensional array. Normally, we actually center the value of k around 0 in defining a Brillouin zone. Thus, to each value of k_x, a distance

$$\Delta k_x = \frac{2\pi}{Na} = \frac{2\pi}{L_x} \tag{5.23}$$

is connected with this value. Hence, if we want to convert the sum to an integral, we must divide the differential distance dk_x by the amount (5.23) to account for that part of the distance that corresponds to each term in the summation. Thus, we make the conversion of the summation to an integration via

$$\sum_{k_x} \to \int \frac{L_x}{2\pi} dk_x. \tag{5.24}$$

In three dimensions, (5.24) can be written as

$$\sum_{k_x, k_y, k_z} \rightarrow \iiint \frac{L_x}{2\pi} \frac{L_y}{2\pi} \frac{L_z}{2\pi} dk_x \, dk_y \, dk_z. \tag{5.25}$$

The quantity

$$\frac{L_x L_y L_z}{(2\pi)^3} = \frac{V}{(2\pi)^3} \tag{5.26}$$

is termed the *density of states* in momentum (wave vector) space. To achieve the density per unit volume, (5.25) and (5.21) are divided by the volume V. Thus, we can write the density per unit volume as

$$n(x, t) = \frac{2}{(2\pi)^2} \iiint f[k, E(k), t] \, dk_x \, dk_y \, dk_z. \tag{5.27}$$

A factor of 2 has been included to account for the fact that each state is spin degenerate.

The energy depends solely on the magnitude of k, rather than on any single directional component. This suggests that the integral (5.27) will be easier to evaluate if we convert to spherical coordinates. With this change, we write (5.27) as

$$n(x, t) = \frac{2}{(2\pi)^2} \iiint f[k, E(k), t] k^2 \sin \vartheta \, d\vartheta \, d\varphi \, dk. \tag{5.28}$$

The Fermi–Dirac function is independent of the two angles, so these integrations can be carried out immediately and yield a factor of 4π. We then make the change of variables for parabolic energy bands as

$$E = \frac{\hbar^2 k^2}{2m^*}, \qquad k \, dk = \frac{m^*}{\hbar^2} dE, \qquad k = \sqrt{\frac{2m^* E}{\hbar^2}}, \tag{5.29}$$

and

$$n = \frac{1}{\pi^2} \frac{m^*}{\hbar^2} \sqrt{\frac{2m^*}{\hbar^2}} \int_0^\infty \frac{\sqrt{E}\, dE}{1 + \exp\left(\dfrac{E - E_F}{k_B T}\right)}. \qquad (5.30)$$

The quantity

$$\rho(E) = \frac{1}{2\pi^2} \left(\frac{2m^*}{\hbar^2}\right)^{3/2} E^{1/2} \qquad (5.31)$$

is termed the *density of states per unit volume per unit energy*, or more typically just the density of states. In (5.29) and (5.30), we have assumed that the bottom of the conduction band is the zero of energy. This is not always the case, and the more correct form of (5.30) is

$$n = \frac{1}{\pi^2} \frac{m^*}{\hbar^2} \sqrt{\frac{2m^*}{\hbar^2}} \int_{E_c}^\infty \frac{\sqrt{E - E_c}\, dE}{1 + \exp\left(\dfrac{E - E_F}{k_B T}\right)}, \qquad (5.30\text{a})$$

where

$$E = E_c + \frac{\hbar^2 k^2}{2m^*} \qquad (5.29\text{a})$$

and

$$\rho_c(E) = \frac{1}{2\pi^2} \left(\frac{2m^*}{\hbar^2}\right)^{3/2} (E - E_c)^{1/2}. \qquad (5.31\text{a})$$

In general, (5.30a) is a complicated integral, which cannot be evaluated in closed form. Instead, we make an approximation, and that is an assumption that the density is not too high. Rather, we assume that the Fermi energy lies in the band gap and is not close to the conduction band edge. This amounts to assuming that the exponential term in (5.30a) is large compared to unity, so that the factor of unity can be ignored. Then, we may rewrite this equation as[3]

$$
n = \frac{1}{2\pi^2}\left(\frac{2m^*}{\hbar^2}\right)^{3/2} e^{-(E_c - E_F)/k_B T} \int_{E_c}^{\infty} (E - E_c)^{1/2} e^{-(E - E_c)/k_B T}\, dE
$$

$$
= \frac{1}{2\pi^2}\left(\frac{2m^* k_B T}{\hbar^2}\right)^{3/2} e^{-(E_c - F_F)/k_B T} \int_{0}^{\infty} x^{1/2} e^{-x}\, dx
$$

(5.32)

where the substitution $x = (E - E_c)/k_B T$ has been made. The approximation is what we call the *nondegenerate* limit — where the Fermi energy is well out of the conduction band. When this approximation does not hold, we are said to be in degenerate conditions. With the nondegenerate approximation, the integral can now be seen to be a gamma function[4]

$$
\Gamma(s) = \int_{0}^{\infty} x^{s-1} e^{-x}\, dx, \qquad \Gamma(3/2) = \sqrt{\pi}/2.
$$

(5.33)

Finally, we arrive at

$$
n = N_c \exp\left(-\frac{E_c - E_F}{k_B T}\right), \qquad N_c = \frac{1}{4}\left(\frac{2m^* k_B T}{\pi \hbar^2}\right)^{3/2}.
$$

(5.34)

The quantity N_c is termed the *effective density of states*. In a material such as GaAs, the effective mass for electrons is $0.067 m_0$, so that the effective density of states is $4.34 \times 10^{17}\ \mathrm{cm}^{-3}$ at 300 K. In general, it is $2.5(m^*/m_0)^{1.5} \times 10^{19}\ \mathrm{cm}^{-3}$ at 300 K.

In the case of Si, there are six equivalent minima of the conduction band, which are located about 85% of the distance to the zone edge along the equivalent [100] directions. Moreover, these minima are ellipsoids of revolution so that the mass in the directions transverse to the [100] is different from that along these directions. These two masses are called the transverse mass ($m_T = 0.19 m_0$) and the longitudinal mass ($m_L = 0.91 m_0$). The six ellipsoids must be combined to give an equivalent density of states effective mass

$$
m_d = 6^{2/3} (m_T^2 m_L)^{1/3} = 1.06 m_0.
$$

(5.35)

The square on m_T arises because there are two transverse directions, while there is only a single longitudinal direction. The exponent on the factor of 6 comes from taking it inside the parentheses in (5.34). Hence, the effective density of states for the conduction band in Si is $2.73 \times 10^{19} \, \text{cm}^{-3}$ at 300 K.

We can proceed in a quite similar manner for the holes in the valence band. There are two differences in the approach. First, we need the probability that an electron is *absent* in the state, or that a hole is present. This probability is simply related to (5.3) as

$$f_h[x, E(k), t] = 1 - f[x, E(k), t]$$

$$= 1 - \frac{1}{1 + \exp\left(\dfrac{E - E_F}{k_B T}\right)} \qquad (5.36)$$

$$= \frac{1}{1 + \exp\left(-\dfrac{E - E_F}{k_B T}\right)}.$$

The second change is that the density of hole states is measured from the *top* of the valence band. Hence, we may adapt (5.31a) to the case of holes as

$$\rho_v(E) = \frac{1}{2\pi^2}\left(\frac{2m^*}{\hbar^2}\right)^{3/2}(E_v - E)^{1/2}, \qquad (5.37)$$

where the effective mass here is that for holes. Thus, we can write the number of holes, which we denote by p, as[3]

$$p = \frac{1}{2\pi^2}\left(\frac{2m^*}{\hbar^2}\right)^{3/2}\int_{-\infty}^{E_v}\frac{(E_v - E)^{1/2}\,dE}{1 + \exp\left(-\dfrac{E - E_F}{k_B T}\right)}. \qquad (5.38)$$

In the case of electrons, it was assumed that the Fermi energy was below the conduction band edge, so that we could keep just the exponential term. A similar approach is taken here, except we assume that the Fermi energy is sufficiently *above* the valence band edge, and again keep only the exponential term. (Again, this is an assumption of a nondegenerate distribution function.) Then, we can rewrite (5.38) as

$$p = \frac{1}{2\pi^2}\left(\frac{2m^*}{\hbar^2}\right)^{3/2}\exp\left(\frac{E_v - E_F}{k_B T}\right)\int_{-\infty}^{E_v}\exp\left(\frac{E - E_v}{k_B T}\right)(E_v - E)^{1/2}\,dE$$

$$= \frac{1}{2\pi^2}\left(\frac{2m^*}{\hbar^2}\right)^{3/2}\exp\left(\frac{E_v - E_F}{k_B T}\right)\int_0^{\infty} x^{1/2}e^{-x}\,dx \qquad (5.39)$$

$$= N_v \exp\left(\frac{E_v - E_F}{k_B T}\right).$$

We have taken the step of introducing the new variable $x = E_v - E$ in the second line, and then recognized the gamma function. The effective density of states for the valence band is

$$N_v = \frac{1}{4}\left(\frac{2m_v k_B T}{\pi\hbar^2}\right)^{3/2}. \qquad (5.40)$$

From Fig. 4.11, it may be ascertained that two bands are degenerate at the top of the valence band at $k = 0$. One of these has high curvature, and therefore a small effective mass. The other has low curvature and therefore a relatively large effective mass. We cleverly call these the light-hole and heavy-hole bands. The total density of states comes from adding these two together, from which we can define

$$m_v^{3/2} = m_{lh}^{3/2} + m_{hh}^{3/2}. \qquad (5.41)$$

This arrangement is correct for almost all semiconductors. In Si, the combined valence band density of states mass m_v is approximately $0.6m_0$. Hence, the effective density of states in the valence band N_v is $1.16 \times 10^{19}\ \mathrm{cm}^{-3}$ at 300 K.

These two results, for the number of electrons and the number of holes, are now combined to account for finding where the Fermi energy has to be, and just how many electrons and holes exist at any temperature. The essential point in determining both of these quantities is that the number of electrons must be equal to the number of holes in order to maintain charge neutrality in the semiconductor. In this intrinsic situation, we write

$$n = p = n_i, \tag{5.42}$$

in which the *intrinsic* concentration (of electrons or holes) n_i is defined. Moreover, we can say that, in equilibrium,

$$n_i^2 = np = N_c N_v \exp\left(\frac{E_v - E_F}{k_B T}\right) \exp\left(-\frac{E_c - E_F}{k_B T}\right)$$

$$= N_c N_v \exp\left(-\frac{E_c - E_v}{k_B T}\right) = N_c N_v \exp\left(-\frac{E_G}{k_B T}\right). \tag{5.43}$$

Here, we have introduced the fundamental energy gap $E_G = E_c - E_v$ as the separation between the conduction and valence bands. From (5.43), we can now determine the intrinsic concentration, which is the equilibrium number of electrons in the conduction band and holes in the valence band. This is 1.54×10^{10} cm^{-3} at 300 K, where the band gap for Si has been taken to be 1.08 eV. However, if we raise the temperature to 600 K, the intrinsic concentration grows to 1.48×10^{15} cm^{-3}. Once again, we see the exponential growth factor. Merely raising the temperature by a factor of 2 has caused an almost five order of magnitude growth in the intrinsic concentration.

Once more using (5.42) will enable us to determine the position of the Fermi energy through a detailed balance requirement between electrons and holes. We can write the left two terms as

$$N_c \exp\left(-\frac{E_c - E_F}{k_B T}\right) = N_v \exp\left(-\frac{E_F - E_v}{k_B T}\right), \tag{5.44}$$

or

$$E_F = \frac{E_c - E_v}{2} + \frac{k_B T}{2} \ln\left(\frac{N_v}{N_c}\right) = \frac{E_G}{2} + \frac{3k_B T}{4} \ln\left(\frac{m_v}{m_d}\right). \tag{5.45}$$

For all practical purposes, the Fermi energy is located at the center of the energy gap between the conduction and valence bands. It is moved slightly away from this point in response to the fact that there is a difference in the number of effective states in the two bands, and it must balance these in a way that equalizes the two densities. For example, at 300 K, the Fermi energy is moved only 11 meV below the center of the gap, and this is only doubled at 600 K. Here, the shift is linear in the temperature, since this occurs in the exponential itself.

5.3. EXTRINSIC SEMICONDUCTORS

Now let us turn to the question of whether we can do something to the semiconductor to create more electrons or holes. In a sense, the semiconductor is rather unique, in that each atom has just enough electrons to satisfy the bond, and the crystal structure is just the right one for which the valence band is completely full and the conduction band is completely empty (at low temperature). It is this fact that makes it possible to control the properties. For example, if we use an impurity, which does not have four electrons, to replace one of the silicon atoms, we will upset this balance. (While the same considerations will apply to the zinc-blende structures, we will deal here primarily with silicon, as the effects are straightforward in this case.) Suppose we replace one of the silicon atoms with a phosphorous atom. Phosphorous has five electrons in its outer shell. As we can see

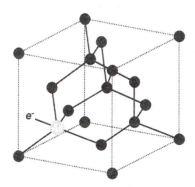

Figure 5.2. A schematic illustration of *n*-type doping. The black symbols denote silicon atoms and we show what happens when an arsenic atom replaces a silicon atom in the crystal structure. An extra electron is associated with the additional, unterminated, bond on the As.

from Fig. 5.2, four of the electrons go to the tetrahedral bonds with the neighboring silicon atoms. But, what about the fifth electron? It is not needed for the tetrahedral bonds. This electron is weakly bonded to the phosphorus atom, so it can easily be excited into an "itinerant" state where it can wander through the lattice. In fact, four of the electrons satisfy the bonds, and these four are just enough to create a completely full valence band. If the fifth electron becomes mobile, it must go into the conduction band, as this is the only available state in the band structure description. The phosphorus atom is termed a "donor," as it provides an extra (donated) electron. The small energy needed to excite this fifth electron is called the donor ionization energy E_d. These energies are typically of the order of 10 meV, so it is clear that they can be thermally excited without much trouble.

On the other hand, suppose we replace a silicon atom with a boron atom. Boron has only three electrons in its outer shell. Thus, it cannot complete the tetrahedral bond. This is shown in Fig. 5.3. As a result, one of the neighboring silicon atoms has a dangling bond electron. But, which one? It could be any of the four neighbors. In fact, this makes it highly likely that the void moves from one bond to the next, even jumping to silicon atoms some distance removed. This process, of course, has an excitation energy, which is of the same order of magnitude as the donor ionization energy. In the band picture, we do not have enough electrons to satisfy all the bonds, so

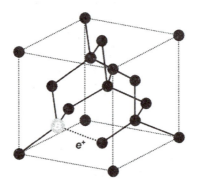

Figure 5.3. A schematic illustration of p-type doping. The black symbols denote silicon atoms and we show what happens when a boron atom replaces a silicon atom in the crystal structure. A hole is associated with the missing bond introduced by the B atom.

that the valence band is not completely full. We recognize that a "hole" has been created within the valence band. The boron atom is termed an "acceptor," as it creates the conditions for the void to move through the various bonds. The energy needed to allow this void to move is called the acceptor ionization energy, and again is of the order of 10 meV.

The presence of donors and acceptors (dopants) changes the statistics of the semiconductor. An extra electron in the conduction band, provided by the phosphorus atom, does not upset the neutrality, as the phosphorus atom itself is positively charged once the electron moves away. Similarly, when the void moves away from the boron atom, the latter becomes negatively charged, which neutralizes the free hole that is traveling through the crystal. Hence, the balance of charge that is expressed by (5.42) must now be written as

$$p + N_D^+ = n + N_A^-, \qquad (5.46)$$

where N_D^+ is the number of ionized donors and N_A^- is the number of "ionized" acceptors. Obviously, if we want a material which has predominantly holes as charge carriers, we place a large number of acceptors in the semiconductor. On the other hand, if we want a material which has predominantly electrons as the charge carrier, we place a large number of donors in the semiconductor. It is not surprising that we term these two cases p-type and n-type semiconductors, respectively. Importantly, we can apply exactly the same procedures of the last section in order to determine the statistics of the doped semiconductor. Here, we will treat the n-type semiconductor, but the p-type follows by a straightforward extension.

In Fig. 5.4, we indicate the energy levels of the donors and acceptors with respect to the conduction and valence bands. The donor level is placed just below the conduction band edge so that $E_c - E_d$ is the donor activation energy — a few meV. Similarly, the acceptor level is placed just above the valence band so that $E_a - E_v$ is the acceptor activaton energy. The donor activation energy is that required to move the excess electron from the donor atom to the conduction band. The acceptor activation energy is that required to move an electron from the valence band to the acceptor — this is the energy to move an electron from one filled bond to the empty bond, allowing the empty bond to move. In n-type material, the number of

Figure 5.4. The variation of the energy band structure of the semiconductor is plotted as a function of position in the crystal. The vertical axis represents energy, while the horizontal axis corresponds to moving in a given direction in the crystal. In addition to plotting the top of the valence band (E_v), and the bottom of the conduction band (E_c), the positions of typical donor (E_d) and acceptor (E_a) levels are also shown in this figure. The differences $E_c - E_d$ and $E_a - E_v$ are the ionization energies.

ionized donors corresponds to the number of donors from which the excess electron has been "stripped." In this sense, an empty state exists at the donor energy level, so that we can say that

$$N_d^+ = N_d[1 - f(E_d)], \qquad (5.47)$$

where $f(E_d)$ is the Fermi–Dirac distribution. We see that we are treating the number of ionized donors precisely as we did states in the valence band. If nearly all the donors are ionized, then we can simplify (5.46) as

$$n \approx N_d^+. \qquad (5.48)$$

We can proceed by equating (5.47) and (5.34) to give[3]

$$N_c \exp\left(-\frac{E_c - E_F}{k_B T}\right) = N_d \frac{1}{1 + \exp\left(\dfrac{E_F - E_d}{k_B T}\right)}. \qquad (5.49)$$

A little algebra will allow us to rearrange the various terms in order to arrive at

$$2 \exp\left(\frac{E_F - E_d}{k_B T}\right) = \sqrt{1 + \frac{N_d}{N_c} \exp\left(\frac{E_c - E_d}{k_B T}\right)} - 1. \qquad (5.50)$$

At low temperatures, $N_d \gg N_c$, the second term in the square root factor is much greater than unity. Under these conditions, we can rewrite (5.50) as

$$\exp\left(\frac{E_F - E_d}{k_B T}\right) \approx \sqrt{\frac{N_d}{4N_c}} \exp\left(\frac{E_c - E_d}{2k_B T}\right), \tag{5.51}$$

and the Fermi energy is given as

$$E_F \approx \frac{E_c + E_d}{2} + \frac{k_B T}{2} \ln\left(\frac{N_d}{4N_c}\right). \tag{5.52}$$

Hence, we see that the donor energy level plays a role similar to the valence band, as at low temperature, the Fermi energy is midway between the conduction band and the donor energy. In this range, the number of electrons is given by inserting (5.52) back into the left-hand side of (5.49), or

$$n = \sqrt{\frac{N_c N_d}{4}} \exp\left(-\frac{E_c - E_d}{2k_B T}\right). \tag{5.53}$$

In this region, not all of the donors are ionized, so that the electron density increases exponentially with a temperature variation given by the pseudo-gap $E_c - E_d$. This region is termed the *reserve* region, as many donors are still neutral. In this region, the donor energy level really does play a similar role to that of the valence band in the intrinsic semiconductor.

At higher temperatures, however, all of the donors are ionized. In this case, we can rewrite (5.49), according to the situation when the Fermi energy lies *below* the donor energy, as

$$n = N_d = N_c \exp\left(-\frac{E_c - E_F}{k_B T}\right). \tag{5.54}$$

In this situation, the number of electrons is constant and given by the number of donor atoms. The Fermi level still moves, however, and this is given by (5.54) to be

$$E_F \approx E_c - k_B T \ln\left(\frac{N_c}{N_d}\right). \tag{5.55}$$

This region is termed the *exhaustion* region, as all donors have been ionized.

In Fig. 5.5, we plot the electron density and the Fermi energy as a function of temperature for these two regions. In addition, we also include the intrinsic region where the temperature rises to a point that electrons are excited from the valence band to the conduction band. Even in the intrinsic region, the number of electrons remains larger than the number of holes. We can see this from (5.46) by ignoring the acceptors, and assuming all the donors are ionized, so that

$$n = N_d^+ + p = N_d^+ + \frac{n_i^2}{n}, \tag{5.56}$$

so that

$$n = \frac{N_d^+}{2} + \sqrt{n_i^2 + \frac{(N_d^+)^2}{4}}, \tag{5.57}$$

and n_i is given by (5.43). In (5.56), we have used the fact that $pn = n_i^2$ is always true in any equilibrium situation. The presence of the ionized donors increases the electron concentration n, and at the same time decreases the hole concentration p to maintain total charge neutrality.

As an example of these statistics, let us consider a situation in which a donor, with an ionization energy of 49 meV, is placed in silicon with a concentration of $2 \times 10^{17}\,\text{cm}^{-3}$. First, we note that when the donors are fully ionized, $n \sim N_d$, and the Fermi energy will be below the donor energy. We note that this concentration is well above the intrinsic concentration at 300 K, so we can use (5.55) to determine just where the Fermi energy lies at this temperature. We find

$$E_F - E_c = -0.0259 \ln\left(\frac{2.73 \times 10^{19}}{2.0 \times 10^{17}}\right) = -127\,\text{meV}. \tag{5.58}$$

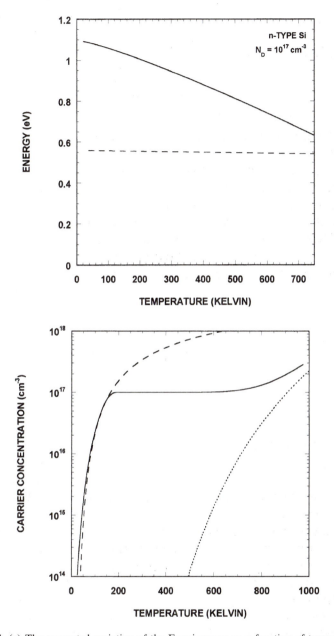

Figure 5.5. (a) The computed variation of the Fermi energy as a function of temperature in *n*-type silicon. The dashed line represents the corresponding variation of the intrinsic Fermi level. Note that at higher temperatures than those shown here the Fermi level in the *n*-type material becomes essentially equal to that of the intrinsic material, corresponding to the intrinsic regime. (b) Variation of the carrier concentration in a doped semiconductor. The dashed and dotted lines correspond to the predictions of Eqs. (5.53) and (5.54), respectively, while the solid line shows the actual variation. At intermediate temperatures, the carrier concentration saturates at a value close to the doping density.

Hence, the Fermi energy lies well below the donor energy. From (5.57), we note that the electron concentration will begin to be intrinsic once $n_i \sim N_d$. When the electron concentration has doubled, we can use (5.57) to find that $n_i = \sqrt{2} N_d^+ \sim 2.8 \times 10^{17}\,\mathrm{cm}^{-3}$ for our current example. Now, we can use (5.43) to find that this occurs for a temperature of the order of 1040 K. So, for most of the important temperature range (~ 300–500 K), modest dopant concentrations will control the properties of the silicon material. *It is this result that makes semiconductor devices possible.*

5.4. ELECTRICAL CONDUCTIVITY

We now want to turn our attention to the current that is carried by the electrons in *n*-type material, or by the holes in *p*-type material, or by both in intrinsic material. In principle, we apply an electric field to the semiconductor, with this field arising from a potential difference between the two ends of the material. In response to this field, the electrons (we treat the electrons for the moment) begin to move according to Newton's law

$$\mathbf{F}_e = m_e \frac{d\mathbf{v}}{dt} = \hbar \frac{d\mathbf{k}}{dt} = -e\mathbf{F}. \tag{5.59}$$

(In order to avoid confusion between the energy and the electric field, we use the symbol **F** for the field. However, this now creates confusion with the force, so we will use the latter only sparingly.) We may integrate (5.59) for a constant field to obtain

$$\mathbf{k}(t) = \mathbf{k}(0) - \frac{e\mathbf{F}}{\hbar} t. \tag{5.60}$$

In this form, the wave vector continuously increases with time. The instantaneous velocity is given by the derivative of the energy relation, or

$$\mathbf{v}(t) = \frac{1}{\hbar} \frac{dE(\mathbf{k})}{d\mathbf{k}}. \tag{5.61}$$

The co-sinusoidal band we found in the last chapter may be written as

$$E(k) = \frac{\Delta}{2}\left[1 - \cos\left(\frac{ka}{\pi}\right)\right], \qquad (5.62)$$

and

$$v(t) = \frac{\Delta a}{2\hbar\pi}\sin\left(\frac{k(t)a}{\pi}\right), \qquad (5.63)$$

in one dimension. This velocity oscillates with time, and its average over many periods is zero. In fact, the electron is traveling through the entire Brillouin zone and the consequent band. In this state, no current can be carried because the average velocity is zero. Something else must affect the movement of the electrons. This something else is scattering.

Scattering occurs because the electron can run into the lattice atoms, which are moving themselves do to their thermal energy. It can also run into impurity atoms, which are ionized and are therefore Coulomb scattering centers. In general, the electron can only go a time τ before, on the average, hitting something and changing its momentum. We can include this in the above discussion, by noting that the probability that the electron has *not* undergone a collision may be expressed as

$$P(t) = \frac{1}{\tau}e^{-t/\tau}. \qquad (5.64)$$

Then, the average wave vector is found by taking the average of the free flight time t, as

$$\langle k \rangle = -\frac{e\mathbf{F}}{\hbar}\int_0^\infty tP(t)\,dt = -\frac{e\mathbf{F}}{\hbar}\int_0^\infty \frac{t}{\tau}e^{-t/\tau}\,dt = -\frac{e\tau}{\hbar}\mathbf{F}. \quad (5.65)$$

Now, we can use our effective mass relationship to find the velocity as

$$\langle v \rangle \equiv \mathbf{v}_{\mathrm{drift}} = \mathbf{v}_d = \frac{\hbar\langle k \rangle}{m_e} = -\frac{e\tau}{m_e}\mathbf{F}. \qquad (5.66)$$

Here, we use the subscript "e" to stand for the electrons; thus, the mass in (5.66) is the electron effective mass. The last expression of (5.66) defines the electron *mobility* as

$$\mu_e = \frac{e\tau}{m_e}. \tag{5.67}$$

In bulk silicon, with modest doping (less than $10^{17}\,\text{cm}^{-3}$), the mobility of the electrons at room temperature is about $1500\,\text{cm}^2/\text{Vs}$. In GaAs, on the other hand, the mobility of electrons is about 7500 cm^2/Vs, due to the much smaller effective mass of the electrons in this material.

In a similar manner, the transport of the holes can be easily developed, using the hole scattering time τ_h and effective mass m_h. Since the holes move in the opposite direction (their sign is reversed), the drift velocity for holes is given by analogy with (5.66) as

$$\langle \mathbf{v} \rangle \equiv \mathbf{v}_{\text{drift},h} = \mathbf{v}_{d,h} = \frac{\hbar \langle \mathbf{k} \rangle}{m_h} = \frac{e\tau_h}{m_h} \mathbf{F}, \tag{5.68}$$

and the hole mobility is

$$\mu_h = \frac{e\tau_h}{m_h}. \tag{5.69}$$

In bulk silicon, the hole mobility is about $400\,\text{cm}^2/\text{Vs}$. This difference is primarily due to the difference in the effective mass. The hole effective mass is mainly that of the heavy holes in silicon. On the other hand, the conductivity effective mass for the electrons is quite different from the density of states effective mass — instead it is given as $0.32m_0$. This difference between the density of states effective mass and the conductivity effective mass lies in the multi-valley nature of the energy band in silicon. The density of states effective mass is heavily weighted by the larger mass, the longitudinal mass, while the conductivity mass is heavily weighted by the smaller mass, the transverse mass. The subtleties of the difference are beyond the present treatment, but one should be aware that this difference exists. There is not such a difference in the valence band, or in the conduction band of materials like GaAs, which has only a single minimum of the conduction band.

In Fig. 5.6, we plot the mobility of electrons and holes in silicon. In panel (a), we plot the mobilities as a function of temperature when there are not a large number of impurities. In panel (b), we plot the mobility as a function of the number of impurities (donors or acceptors) at room temperature. At low temperature, the mobility increases with temperature, as the scattering is dominated by Coulomb scattering of the carriers by the ionized impurities (inset to Fig. 5.6(a)). At higher temperatures, the mobility decreases with temperature as the scattering is dominated by lattice vibrations (phonons). In general, the mobility increases as $T^{3/2}$ at low temperature — the scattering time τ decreases as $T^{-3/2}$. At high temperatures, the power law is more complicated as there are both short range and long range lattice vibrations that interact with the carriers. As a general rule, however, the mobility of the electrons decreases as $T^{-1.66}$ in the higher temperature regime. The mobility of the holes is quite similar. The dependence on the impurity concentration is due to the larger number of Coulomb scattering centers as the dopant concentration is increased. The mobility in the low temperature regime is reduced, which means that the peak in Fig. 5.6(a) moves to the right. At a sufficiently high dopant concentration, the mobility begins to decrease due to this greater number of ionized impurities. The combined effect shown in Fig. 5.6(a) implies that we can add scattering *rates* as

$$\frac{1}{\tau} = \frac{1}{\tau}\bigg|_{\text{impurity}} + \frac{1}{\tau}\bigg|_{\text{lattice}}. \tag{5.70}$$

This additive process is known as Mattheisen's rule, and implies that each scattering process is independent of any other scattering process. Hence, the effects of each can be added to a common sum.

We can now compute the current density in the semiconductor. For electrons, the current density is the product of the number of charges and their velocities, or[5]

$$\mathbf{J}_e = -ne\mathbf{v}_d = ne\mu_e\mathbf{F} = \frac{ne^2\tau_e}{m_e}\mathbf{F} = \sigma_e\mathbf{F}. \tag{5.71}$$

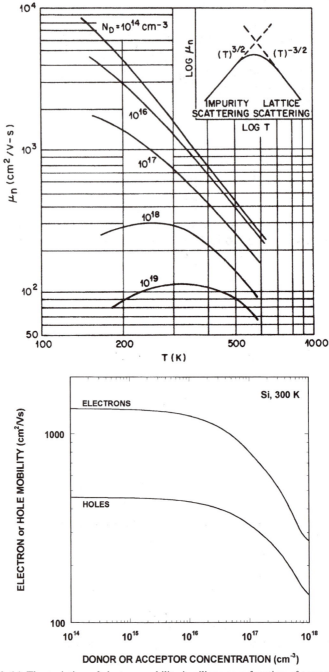

Figure 5.6. (a) The variation of electron mobility in silicon as a function of temperature and doping density. From "Semiconductor Devices, Physics and Technology," by S. M. Sze. *Copyright © John Wiley & Sons, 1985.* Reprinted by permission of John Wiley & Sons, Inc. (b) The variation of the electron and hole mobilities in silicon as a function of the doping concentration. The temperature is held constant here at 300 K.

The last expression introduces the *conductivity* σ_e for electrons. In essence, (5.71) is Ohm's law (the point form). If we integrate over the transverse area, and over the length, we introduce the area and length to (5.71). For homogeneous material, we can rewrite (5.71) as (in scalar form)

$$\frac{I_e}{A} = \sigma_e L \frac{V}{L}, \qquad R = \frac{L}{\sigma_e A} = \frac{\rho_e L}{A}, \tag{5.72}$$

where $\rho_e = 1/\sigma_e$ is the *resistivity* of the electrons. By a similar treatment, we can introduce the hole conductivity as

$$\mathbf{J}_h = pe\mathbf{v}_{dh} = pe\mu_h\mathbf{F} = \frac{pe^2\tau_h}{m_h}\mathbf{F} = \sigma_h\mathbf{F}. \tag{5.73}$$

The total current density is the sum of the contributions from the electrons and the holes, and may be expressed as

$$\mathbf{J} = \mathbf{J}_e + \mathbf{J}_h = (\sigma_e + \sigma_h)\mathbf{F} = \sigma\mathbf{F}. \tag{5.74}$$

In Fig. 5.7, we plot the resistivity of *n*-type silicon as a function of temperature. At low temperatures, the resistivity drops because the mobility is increasing as a function of temperature and the number of free carriers is increasing as more donors are ionized. In the intermediate temperature range, the number of free carriers is constant, as all the donors are ionized, but the mobility may be either increasing or decreasing, depending on the number of ionized donors (that is, whether the mobility is dominated by the impurities or by lattice scattering). At the highest temperatures, the carrier concentration is increasing exponentially — the intrinsic regime — and this overwhelms the decreasing mobility. This rapid decrease of resistivity was mentioned in Chapter 1 as the tell-tale sign by which Faraday knew that he had discovered a new type of material (in his case, he was studying AgS).

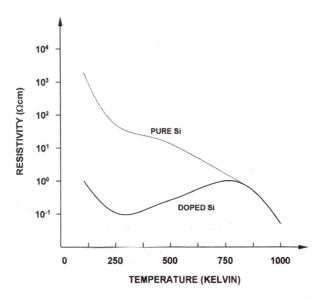

Figure 5.7. Schematic diagram illustrating typical variations of the resistivity of silicon samples as a function of temperature. The various regions are discussed in the text. However, the rapid decrease in resistivity with temperature, in the high temperature regime, is a tell-tale sign of a semiconductor. This is the intrinsic region, where the exponential increase of carrier concentration overwhelms any decreasing mobility.

5.5. CONDUCTIVITY IN A MAGNETIC FIELD

When we apply a magnetic field to the semiconductor, as well as an electric field, it becomes necessary to consider the vector nature of the forces and motion. That is because the magnetic field induced forces are normal to the magnetic field itself. They are also velocity dependent, but normal to the velocity itself. In the previous section, vector relations were used, but these were for convenience, and not a necessity. Here, they will be a necessity. In the presence of both electric and magnetic fields, the force provides the driving terms in Newton's second law of motion. For the electrons, we write this as[6]

$$m_e \frac{d\mathbf{v}_d}{dt} = -e(\mathbf{F} + \mathbf{v}_d \times \mathbf{B}), \qquad (5.75)$$

where the term on the right-hand side of the equation is the *Lorentz force*.

5.5.1. Low Magnetic Field

For the present, we will assume a steady state exists with the time variation of the velocity given by collisions, so that we may rewrite (5.75) as

$$m_e \frac{\mathbf{v}_d}{\tau_e} = -e(\mathbf{F} + \mathbf{v}_d \times \mathbf{B}). \tag{5.76}$$

Let us consider the situation in Fig. 5.8, in which a semiconductor is assumed to be aligned to the normal cartesian coordinates. We will assume that the magnetic field is in the z direction, and that the current and electric field are in the x direction. That is, the electric field is applied in the x direction, and we will force a *constraint* onto the solution which says that the current is only in the x direction. With this orientation of the fields, (5.76) can be rewritten as (motion in the z direction is ignored)

$$v_{dx} = -\mu_e(F_x + v_{dy}B_z),$$
$$v_{dy} = -\mu_e(F_y - v_{dx}B_z). \tag{5.77}$$

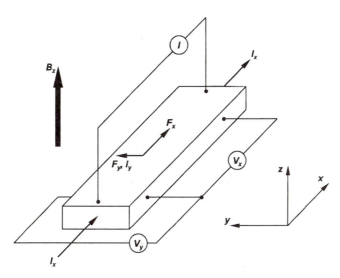

Figure 5.8. The experimental configuration of a Hall-effect measurement of a semiconductor is schematically illustrated. Note that in experiment, the magnetic field is applied perpendicular to the plane of the sample, and so is perpendicular to the measured current and the voltages.

These two equations are for the velocity in the plane normal to the magnetic field. Since we are dealing only with electrons at the present time, we may force the constraint of zero current in the y direction onto the problem by setting $v_{dy} = 0$. Hence, the second of equations (5.77) may be rewritten as

$$F_y = v_{dx} B_z. \tag{5.78}$$

The same approximation applied to the first equation of (5.77) says that the x directed velocity is unaffected by the magnetic field. This is true only for very small magnetic fields—just how small will be discussed below. However, we can use this fact to rewrite (5.78) as

$$F_y = v_{dx} B_z = -\mu_e F_x B_z = -\frac{J_x}{ne} B_z. \tag{5.79}$$

The *Hall coefficient* R_H is defined through the equation

$$F_y = R_H J_x B_z, \tag{5.80}$$

so that, for electrons, the Hall coefficient is given by

$$R_H = -\frac{1}{ne}. \tag{5.81}$$

This overall effect is called the *Hall effect*.[7] The result of passing a current through the semiconductor, normal to a magnetic field, results in an induced transverse electric field which is normal to both the current and the magnetic field. In the present case, this Hall field is F_y. The size of the field is given directly by the current and field and the Hall coefficient. Measuring the Hall coefficient is the most common method of determining just how many electrons are present in the semiconductor. For example, if we have 10^{17} electrons per cm^3, then the Hall coefficient is 62.5 cm^3/Coulomb. If we pass 1 mA/cm^2 through the semiconductor, in the presence of a 0.1 T magnetic field, the transverse electric field is 6.25×10^{-7} V/cm, and is directed in the negative y direction. Measuring the direction of the field tells us whether we are dealing with electrons or holes.

Let us now turn to the case where both electrons and holes are present. We may write the equivalent equations to (5.77) for holes as[8]

$$v_{hx} = \mu_h(F_x + v_{hy}B_z),$$
$$v_{hy} = \mu_h(F_y - v_{hx}B_z). \tag{5.82}$$

Before we can apply any constraints, however, we need to put these equations together with (5.77) to get the total currents in each direction, which are

$$J_x = J_{ex} + J_{hx} = (pe\mu_h + ne\mu_e)F_x + (pe\mu_h v_{hy} + ne\mu_e v_{ey})B_z$$
$$\approx (pe\mu_h + ne\mu_e)F_x, \tag{5.83}$$
$$J_y = J_{ey} + J_{hy} = (pe\mu_h + ne\mu_e)F_y - (pe\mu_h v_{hx} + ne\mu_e v_{ex})B_z.$$

In the first equation above, we have assumed that the magnetic field is sufficiently small that we can ignore the second term. Now, we can impose our constraint that the total current in the y direction vanishes, so that

$$F_y = \frac{pe\mu_h v_{hx} + ne\mu_e v_{dx}}{pe\mu_h + ne\mu_e} B_z = \frac{pe\mu_h^2 - ne\mu_e^2}{pe\mu_h + ne\mu_e} F_x B_z = \frac{pe\mu_h^2 - ne\mu_e^2}{(pe\mu_h + ne\mu_e)^2} J_x B_z.$$

$$\tag{5.84}$$

Comparing this with (5.80) immediately gives the Hall coefficient to be

$$R_H = \frac{pe\mu_h^2 - ne\mu_e^2}{(pe\mu_h + ne\mu_e)^2}. \tag{5.85}$$

If there are no holes, then (5.85) becomes just (5.81) as expected. On the other hand, if there are no electrons, then the Hall coefficient for the holes is

$$R_{H,h} = \frac{1}{pe}. \tag{5.86}$$

Thus, the sign of the Hall coefficient, and the consequent transverse field, is different for electrons and holes. In the case where both carriers are present, we can simplify (5.85) by introducing the mobility ratio $b = \mu_e/\mu_h$. Then, (5.85) can be written as

$$R_H = -\frac{1}{e}\frac{b^2n - p}{(bn + p)^2}. \qquad (5.87)$$

One might expect the Hall coefficient to vanish when the numbers of electrons and holes are equal, but this is not the case. The fact that the two types of carriers have different mobilities means that they respond to the Lorentz force with different transverse deflections. Thus, the relative concentrations at which the Hall coefficient vanishes depends upon these mobilities. As a result the Hall coefficient vanishes when

$$R_H \to 0 \quad \text{for } b^2n = p. \qquad (5.88)$$

Since the mobility of electrons is usually higher than that for holes, the hall coefficient vanishes in p-type material when it begins to approach the intrinsic regime, as a general rule. In silicon, for example, using the values of mobility given above, this vanishing occurs for $n \sim 0.07p$.

5.5.2. High Magnetic Field

In the preceding, we assumed that the magnetic field was small, and that the major component of the time variation came from the scattering time. Here, we want to make the opposite approximation — that is, we will ignore the scattering time in this section. Hence, we return to equation (5.75), which gives the two equations, in the plane normal to the magnetic field, in the absence of an electric field[9]

$$\frac{dv_x}{dt} = -\frac{e}{m_e}v_y B_z,$$

$$\frac{dv_y}{dt} = \frac{e}{m_e}v_x B_z. \qquad (5.89)$$

We can now eliminate one velocity from the equation for the other to obtain a second-order differential equation, as

$$\frac{d^2 v_x}{dt^2} = -\frac{e}{m_e}\frac{dv_y}{dt}B_z = -\left(\frac{eB_z}{m_e}\right)^2 v_x. \qquad (5.90)$$

This can be solved to give

$$v_x = v_0 \cos(\omega_c t), \qquad \omega_c = \frac{eB_z}{m_e}. \qquad (5.91)$$

The last expression defines the *cyclotron frequency* ω_c. In a similar manner, the other velocity component can be found to be

$$v_y = -v_0 \sin(\omega_c t). \qquad (5.92)$$

Here, v_0 is the single unknown constant that arises from the differential equation (5.90). The second constant has been used in assuring that the total velocity is in the x direction at $t = 0$.

From equations (5.91) and (5.92), we can find the position of the particle at any time t by integrating these equations. This gives

$$x = \frac{v_0}{\omega_c}\sin(\omega_c t), \qquad y = \frac{v_0}{\omega_c}\cos(\omega_c t). \qquad (5.93)$$

From this result, it is clear that we can go over to cylindrical coordinates, and the electron describes a circular motion with constant radius

$$r = \sqrt{x^2 + y^2} = \frac{v_0}{\omega_c}. \qquad (5.94)$$

At the same time, the energy of this electron is given by

$$E = \tfrac{1}{2}m_e v^2 = \tfrac{1}{2}m_e v_0^2. \qquad (5.95)$$

Hence, the unknown parameter v_0 is set by the energy of the electron.

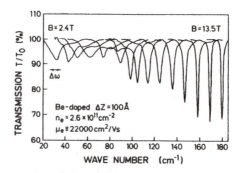

Figure 5.9. Measurements of the microwave absorption of a two-dimensional electron gas, localized at the interface of a GaAs–AlGaAs heterojunction, are shown as a function of magnetic field. The various minima are for different values of the magnetic field, ranging from 2.4 T, on the left, to 13.5 T, on the right. The increase in the wave number of the absorption dip is linearly proportional to the magnetic field. (Here, the wave number $k = 2\pi/\lambda - 2\pi f/c$, where f is the microwave frequency. *J. Richter et al., Phys. Rev. B* **39,** *6268 (1989). Figure reproduced with the permission of the authors.*

With a little intuition, we can now define the regions of low and high magnetic field. When the magnetic field is low, the scattering is sufficiently rapid to break up this circular motion of the electron — the cyclotron motion. On the other hand, when the magnetic field is high, the electron can make several orbits before undergoing a collision. Thus, it is the ratio of the cyclotron frequency to the collision frequency that describes the border between low magnetic field and high magnetic field behavior. That is, for low magnetic fields, we generally have $\omega_c \tau < 1$, while for high magnetic fields, we have the opposite situation $\omega_c \tau > 1$. We can then say that, for low magnetic fields, we should predominantly see the Hall effect. On the other hand, for high magnetic fields, we will observe the orbiting properties of the electron. This can be seen quite clearly if we illuminate the semiconductor sample with microwave radiation. When the microwave frequency is equal to the cyclotron frequency, the electron will absorb energy from the microwaves. Hence, at this "resonance," the electron gas will absorb energy, and the transmission through the gas will show a dip. This is shown in Fig. 5.9 for a two-dimensional electron system at the interface of the GaAs–AlGaAs heterostructure. A linear increase, in the frequency at which the absorption occurs, arises from the increase in the magnetic field.

5.5.3. The Quantum Hall Effect

The motion of the electron in its circular orbit about the magnetic field lines has all the characteristics of a harmonic oscillator. In the previous section, this was treated in the classical manner, but how do the quantum properties arise? The answer lies in the de Broglie wavelength that has been so essential in the previous chapters. Once more, it gives us the basics of the quantization process. In fact, de Broglie first arrived at his wavelengths in terms of electrons orbiting the nucleus of an atom. Here, we have electrons orbiting around magnetic field lines. While the two systems are totally different in detail, the physics of the quantization is quite the same. The ideas of de Broglie tell us that each orbit must contain an integer number of wavelengths of the electron, where the wavelength is given by the wave vector (momentum) of the electron. In fact, this integer must be an odd integer so that the wave function is antisymmetric, as required for electrons (which satisfy the Pauli exclusion principle). Hence, quantization of the electron cyclotron orbits arises from requiring that an odd number of wavelengths fit into the orbit circumference, as

$$2\pi r = (2n + 1)\lambda = (2n + 1)\frac{h}{p} = \frac{2\pi h}{(2n + 1)m_e v_0}. \tag{5.96}$$

Using (5.94), the above equation can be rearranged to yield

$$m_e v_0^2 = (2n + 1)\hbar\omega_c, \qquad E = \tfrac{1}{2}m_e v_0^2 = (n + \tfrac{1}{2})\hbar\omega_c. \tag{5.97}$$

The quantization tells us that the energy levels are separated by the energy of the cyclotron frequency, $\Delta E = \hbar\omega_c$, and thus form a ladder of energy levels. The classical states are pulled together into these discrete levels, which are termed Landau levels, after the Russian physicist Lev Landau, who first worked out these details.

The radius of the orbits is also quantized, since the energy and velocity v_0 are quantized. The radius is given by

$$r_n = \frac{v_0}{\omega_c} = \sqrt{(2n + 1)\frac{\hbar}{m_e \omega_c}} = \sqrt{(2n + 1)\frac{\hbar}{eB_z}}. \tag{5.98}$$

We term the radius of the lowest energy level ($n = 0$) the *magnetic length*, which is

$$l_B = \sqrt{\frac{\hbar}{eB_z}}. \tag{5.99}$$

As the magnetic field is increased, the orbits get smaller, and the Landau level energy increases — confinement of the electrons in this pseudo-potential well costs energy. However, we can also write the radius of the topmost energy level (the one at the Fermi energy level) as

$$r_F = \frac{v_{0,F}}{\omega_c} = \frac{\hbar k_F}{m_e \omega_c} = \frac{\hbar k_F}{eB_z} = k_F l_B^2. \tag{5.100}$$

This tells us that the radius at the Fermi energy is much larger than the magnetic length itself. Rather, comparison of (5.98) and (5.100) leads us to

$$k_F l_B = \sqrt{2n_{\max} + 1}, \tag{5.101}$$

where $n_{\max}$ is the highest occupied Landau level index.

Now, what happens when we make a Hall effect measurement? The general layout of the experiment was already shown in Fig. 5.8. For our considerations, we will assume that the material of interest is a heterostructure, so that the electron gas is approximately a two-dimensional one in the plane of the GaAs–AlGaAs heterostructure. This serves only to constrain the motion in the direction parallel to the magnetic field, which is normal to the heterostructure interface. We will discuss relaxing this constraint later. In the bulk of the semiconductor, away from the edges in the y direction, the energy levels are quantized into Landau levels. As we near the edge of the test structure, however, electrons are repelled from the boundaries. This is introduced by raising of the energy levels as we approach the boundaries, as indicated in Fig. 5.10. The Landau levels bend upward as the boundary is approached, with this bending indicating a surface electric field that prevents the electrons from reaching the edge of the sample. Now, we see an effect that depends on the Fermi energy. If

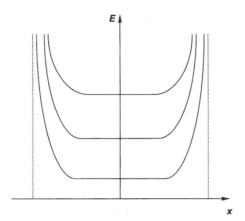

Figure 5.10. Physical boundaries modify the energy levels within a semiconductor. The boundaries are indicated by the dotted lines and cause the Landau levels of the semiconductor to diverge upward as they are approached. This effect can be understood to arise, for example, from repulsion of electrons from the boundaries due to the work function.

the Fermi energy lies in one of the bulk Landau levels, then there are plenty of states which can carry current through the sample. In this case, the conductance of the sample is large. On the other hand, if the Fermi energy lies between two bulk Landau levels, then there are no bulk states to carry current; only a few states at the edges carry this current. In this case, the conductance is very low, *essentially to the point of being zero.* Thus, as we vary the magnetic field, Landau levels are pushed through the Fermi energy as their separation is increased (the Fermi energy's level is fixed by the metallic contacts), and the conductance oscillates. This is shown in Fig. 5.11. This effect is called the Shubnikov–de Haas effect. We can obtain the oscillation period in the *inverse* of the magnetic field from (5.101). This can be rewritten as

$$n_{max} = \frac{k_F^2 l_B^2}{2} - 1 = \frac{\hbar k_F^2}{2eB_z} - 1, \tag{5.102}$$

and

$$\Delta n_{max} = \frac{\hbar k_F^2}{2e} \Delta \left(\frac{1}{B_z} \right) = 1, \tag{5.103}$$

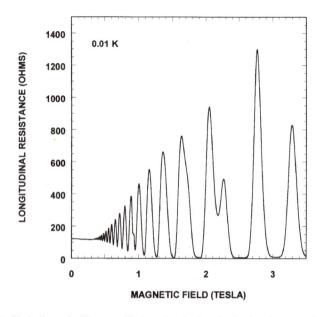

Figure 5.11. Shubnikov–de Haas oscillations in the longitudinal resistance of a semiconductor sample, caused by the emptying of successive Landau levels that are pushed up past the
Fermi level as the magnetic field is increased. These measurements were performed on a high
mobility ($> 10^6$ cm^2/V s) two-dimensional electron gas of a GaAs–AlGaAs heterostructure at
a temperature of 0.01 K. Note how the spacing of successive oscillations *increases* with
increasing magnetic field, consistent with the expected $1/B$ periodicity. Also note how at
magnetic fields in excess of 2 T, the oscillations split into two components as the magnetic field
induces spin splitting. Application of Eq. (5.106) to these oscillations yields a carrier density of
3.8×10^{11} cm^{-2}.

or

$$\Delta\left(\frac{1}{B_z}\right) = \frac{2e}{\hbar k_F^2}. \tag{5.104}$$

Now, in two dimensions at low temperaures, we can approximate
(5.2) as

$$n_s = \frac{1}{2\pi^2}\int d\varphi \int k\,dk\, \frac{1}{1+\exp\left(\dfrac{E(k)-E_F}{k_B T}\right)} \approx \frac{1}{\pi}\int_0^{k_F} k\,dk = \frac{1}{2\pi}k_F^2. \tag{5.105}$$

Hence, we can rewrite (5.104) as

$$\Delta\left(\frac{1}{B_z}\right) = \frac{e}{\pi \hbar n_s}.$$ (5.106)

Here, n_s is the two-dimensional carrier density in the inversion layer at the GaAs–AlGaAs interface. We note that the oscillation is periodic in $1/B_z$, so care must be taken in evaluating (5.106). Nevertheless, this result gives us another method of determining the carrier density from a high magnetic field measurement. One work of caution must be made, however. The density in (5.105) assumes that the levels are spin degenerate. In high magnetic fields, the spin levels begin to be resolved. In this latter case, an extra factor of 2 appears in the denominator of (5.106).

Let us now assume that the number of spin resolved Landau levels is given by the number ν. The density is given approximately by the relation

$$n_s = \frac{\nu}{2}\frac{1}{\pi l_B^2}.$$ (5.107)

We now use this in the defining equation for the Hall coefficient (5.80), and the Hall resistivity becomes

$$\rho_H = \frac{E_y}{J_x} = -\frac{B_z}{n_s e} = -\frac{B_z}{e}\frac{2\pi d_B^2}{\nu} = -\frac{h}{\nu e^2}.$$ (5.108)

Hence, in the regions where the longitudinal conductance is nearly zero, the transverse (Hall) resistance is essentially quantized at a value given by h/e^2 divided by an integer! This was first discovered by Klaus von Klitzing,[10] and is known as the quantum Hall effect. The factor $h/e^2 = 25{,}812\,\Omega$ is the standard of resistance now in most countries (the value is actually known to some seven significant digits). In Fig. 5.12, we plot the Hall resistance for the same sample as shown in Fig. 5.10. It is clear that the plateaus are clear and give very precise values of resistance. Von Klitzing was awarded the Nobel prize for this discovery.

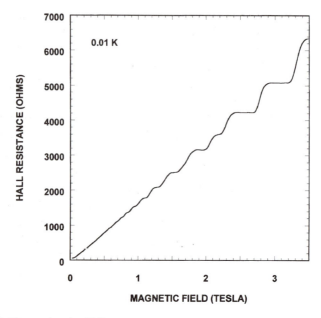

Figure 5.12. The results of a Hall measurement of the sample of Fig. 5.12 at temperature of 0.01 K. The plateaus seen at higher magnetic fields in this figure correspond precisely to the regions where the longitudinal resistance vanishes in Fig. 5.11 and are the plateaus of the quantum Hall effect.

5.6. MAJORITY AND MINORITY CARRIERS

In Sec. 5.2, we developed the intrinsic concentration and the important relation $pn = n_i^2$, given in (5.43). The intrinsic concentration grows exponentially with the temperature, but the product remains valid in *equilibrium*. When donors are added to the semiconductor, the free electron density n increases significantly, but (5.43) remains valid, as we have not caused the semiconductor to depart from equilibrium. Consequently, the number of holes p must decrease as donors are added. This fact was used in (5.56), where it was found that $p \ll n$ in this donor-rich case. Here, we refer to the electrons as the *majority* carriers and the holes as the *minority* carriers. If, by some means, we drive the semiconductor out of equilibrium, then we expect (5.43) to be violated. The nonequilibrium state is generally achieved by the creation of *excess* carriers; that is, we create more carriers than are expected for the equilibrium situation.

Consider a simple case in which we shine light on the semiconductor. If the light has photon energy larger than the band gap $hf > E_G$, then electron–hole pairs can be created. In this process, an electron in the valence band can absorb the photon and jump to the conduction band, leaving a hole behind. We will study this process in some detail in Chapter 6 for device applications. Here, we want to look at the excess carriers that are created. In this optical generation, the number of excess carriers is equal to the number of electron–hole pairs that are created by the photons. A majority of the light is absorbed right at the surface, and the number of photogenerated carriers decreases away from the surface, since the light intensity decreases away from the surface (as photons are absorbed, there remain fewer to be absorbed deeper in the semiconductor). The absorption process creates extra electrons Δn and hole Δp, with $\Delta n = \Delta p$. Under these conditions, (5.43) is violated, and

$$np = (n_0 + \Delta n)(p_0 + \Delta p) = n_i^2 + \Delta n(n_0 + p_0) + \Delta n \Delta p > n_i^2. \quad (5.109)$$

Here, we have identified n_0 and p_0 as the thermal equilibrium values, which satisfy (5.43).

Let us consider that the light is shone on the semiconductor at a surface, which we take as $y = 0$ (the semiconductor is in the region $y > 0$). The excess electron (and hole) concentration decreases away from the surface, as fewer carriers are created in the interior region. This is shown in Fig. 5.13. The spatial variation of the density leads to a particle flow away from the surface—the excess electrons will move toward a region of lower density. Thus, we can say that the velocity of electrons is proportional to the derivative of the density, as

$$v_e \sim -\frac{dn}{dy} = -\frac{d\Delta n}{dy}, \quad (5.110)$$

since the motion is away from the surface. The proportionality constant, once we normalize out the electron density, is called the *diffusion constant* D_e. The proper form for (5.110) is then

$$v_e n \sim -D_e \frac{dn}{dy} = -D_e \frac{d\Delta n}{dy}. \quad (5.111)$$

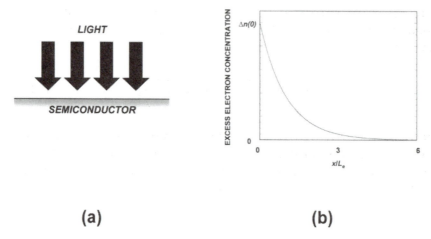

(a)   **(b)**

Figure 5.13. (a) A schematic illustrations of an experiment where light is shone uniformly onto the surface of a semiconductor. (b) The illumination gives rise to an excess electron concentration that decays exponentially as a function of depth from the illuminated surface of the semiconductor.

The units of D_e are cm^2/s (or m^2/s). Of course, the excess holes also peak at the surface, and their density is the same as that of the electrons in Fig. 5.13(b). Thus, the motion of the holes is in the same direction as that of the electrons, and

$$v_h p \sim -D_h \frac{dp}{dy} = -D_h \frac{d\Delta p}{dy}. \tag{5.112}$$

The current carried by these particles, however, is in opposite directions since electrons and holes carry opposite charge. Thus, the total diffusion current is

$$J = -nev_e + pev_h = eD_e \frac{d\Delta n}{dy} - eD_h \frac{d\Delta p}{dy}. \tag{5.113}$$

Suppose, however, that no current is allowed to flow in the semiconductor (for example, there may be no contacts to the material). Then, a voltage must develop between the front surface and the back surface that counteracts this diffusion current. The voltage is a function of position y, and we can in general write the hole concentration as

$$p \sim \exp\left(\frac{E_v - E_F - eV(y)}{k_B T}\right). \qquad (5.114)$$

Normally, the equality would be established with N_v, but this is not the case with the nonequilibrium illumination of the front surface. We can now determine the spatial derivative of the hole concentration as

$$\frac{dp}{dy} = -\frac{pe}{k_B T}\frac{dV(y)}{dy} = \frac{pe}{k_B T}F_y. \qquad (5.115)$$

If we ask that this drift current field produce a current that balances the diffusion current, we need to have

$$J_h = pe\mu_h F_y - epD_h\frac{d\Delta p}{dy} = pe\mu_h F_y - eD_h\frac{pe}{k_B T}F_y = 0, \quad (5.116)$$

or

$$D_h = \frac{\mu_h k_B T}{e}. \qquad (5.117)$$

In a similar manner, we find that

$$D_e = \frac{\mu_e k_B T}{e}. \qquad (5.118)$$

Equations (5.117) and (5.118) are known as Einstein relations. The mobility arises from dissipation in the semiconductor, as characterized by the scattering time τ. On the other hand, the diffusion constant tells about the decay of excess carriers, or fluctuations in the density. Thus, these equations are said to examples of fluctuation–dissipation theorems. It is important to note, however, that non-degenerate statistics were used in (5.114), so that different forms of

these relations will result from degenerate situations where the full
Fermi–Dirac distribution is used.

5.7. LIFETIMES, RECOMBINATION, AND THE DIFFUSION EQUATION

Left to themselves, the excess carriers that have been generated by
the photo-illumination would never go away. They could, of course,
flow out of the ends of the samples, but if current flows, then leaving
carriers are replenished with new arriving carriers in order to
maintain current continuity. In fact, the reverse process of carrier
generation by the photons is carrier *recombination*. We show this
generation and recombination in Fig. 5.14. In this process, an
electron and a hole come together and the electron recrosses the
band gap to the valence band—this eliminates the electron and the
hole. However, this is a very rare process in Si, due to the indirect
band gap, so that the recombination is usually accomplished by a
series of transitions involving impurities and/or defects and phonons
(lattice vibrations). Nevertheless, the excess carriers gradually recombine once the light is turned off, and Δn, $\Delta p \to 0$.

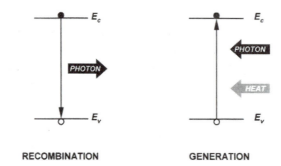

Figure 5.14. Schematic illustration of recombination and generation processes in a semiconductor. In *recombination*, an electron drops from the conduction band to recombine with a hole,
liberating its excess energy in the form of a photon. In *generation*, energy is supplied to the
crystal to excite an electron into the conduction band, leaving a hole behind in the valence
band.

This process of recombination has a characteristic lifetime τ_n (we will use the subscripts n and p for recombination processes, whereas we used subscripts e and h for electron and hole scattering times) for electrons and τ_p for holes. With a definition of a lifetime, we can in general say that the time rate of change of the density is given by

$$\frac{dp}{dt} = \frac{d\Delta p}{dt} = G - R = G - \frac{p}{\tau_p}. \tag{5.119}$$

The quantity G is the generation rate and the quantity R is the recombination rate. The generation process creates excess carriers, while the recombination process gets rid of excess carriers. In the equilibrium case, the population has no external generation, and the population is constant, so that the right-hand side of (5.119) has to vanish. Hence, in equilibrium,

$$G_0 - \frac{p_0}{\tau_p} = 0. \tag{5.120}$$

Here, the subscripts "0" indicate the equilibrium situation. With this relation, we can rewrite (5.119) in terms of the excess generation g as

$$\frac{d\Delta p}{dt} = G_0 + g - \frac{p}{\tau_p} = g - \frac{\Delta p}{\tau_p}. \tag{5.121}$$

Here, we have set the total generation to $G = G_0 + g$, and the excess hole concentration is $\Delta p = p - p_0$.

If we assume that the excess generation is turned "on" at $t = 0$, then (5.121) can be solved to yield the time-dependent excess density as[3]

$$\Delta p(t) = g\tau_p(1 - e^{-t/\tau_p}). \tag{5.122}$$

This result would be the primary result if the illumination was uniform throughout the semiconductor. However, we pointed out in the last section that the generated particles diffused away from the surface, so that the density decreased exponentially away from the surface. Hence, we need to add this term to (5.121) through the continuity equation

$$\frac{\partial \Delta p}{\partial t} = g - \frac{\Delta p}{\tau_p} - \frac{1}{e} \frac{\partial J_h}{\partial y}\bigg|_{\text{diffusion}} = g - \frac{\Delta p}{\tau_p} + D_h \frac{\partial^2 \Delta p}{\partial y^2}. \quad (5.123)$$

Here, the total derivative with respect to time has been changed to a partial derivative in recognition of the y variation, and (5.113) has been used for the current dependence of the holes. In the absence of excess generation, (5.123) is the diffusion equation

$$D_h \frac{\partial^2 \Delta p}{\partial y^2} - \frac{\Delta p}{\tau_p} = \frac{\partial \Delta p}{\partial t}. \quad (5.124)$$

If we assume that all the excess generation occurs at $y = 0$, then it can be treated as a boundary condition on the excess density, and (5.124) is valid within the entire region $y > 0$ (or $y < 0$ for that matter). In this case, one usually assumes a steady-state situation, so that it is the left-hand side of (5.123) that is set to zero, and the excess generation is treated as a boundary condition. Then, the reduced equation that results comes from setting the right-hand side of (5.124) to zero, so that

$$D_h \frac{\partial^2 \Delta p}{\partial y^2} - \frac{\Delta p}{\tau_p} = 0. \quad (5.125)$$

We note that the quantity

$$L_h = \sqrt{D_h \tau_p} \quad (5.126)$$

is a characteristic length for this equation. This length is called the *diffusion length*. In Fig. 5.15, we plot the minority carrier diffusion constant as a function of the doping concentration. In Fig. 5.16, we plot the carrier lifetimes. Indeed, the general solution of (5.125) is of the form $e^{\pm y/L_h}$, so that we can write a solution as

$$\Delta p(y) = A e^{-y/L_h} + B e^{y/L_h}. \quad (5.127)$$

Our problem, however, has solutions in the range $y > 0$, and we require the solution to be bounded (infinite values of the excess

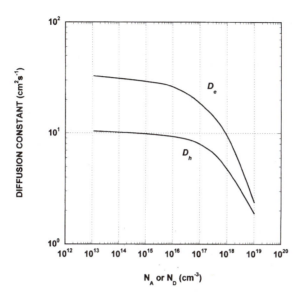

Figure 5.15. Variation of the electron and hole diffusion constants as a function of dopant concentration (N_A or N_D) in n-type and p-type silicon at 300 K.

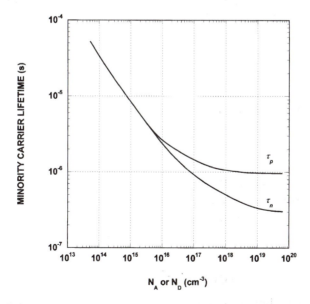

Figure 5.16. Variation of the recombination lifetimes of minority carriers as a function of dopant concentration (N_A or N_D) in n-type and p-type silicon at 300 K.

density are not allowed). Hence, we must set $B = 0$. Then, the excess density is

$$\Delta p(y) = \Delta p(0)e^{-y/L_h}. \tag{5.128}$$

Hence, the density decays in the manner that we assumed earlier.

We can examine the situation discussed in this section and the previous one, where we illuminate the semiconductor on one surface with light. This creates excess carriers, which diffuse away from the surface until a space-charge field is sufficiently large to oppose further diffusion. Since, the total hole current in the y direction is assumed to be zero, we can measure the resulting voltage. Using (5.116), the induced electric field is given by

$$F_y = \frac{D_h}{p\mu_h} \frac{\partial \Delta p(y)}{\partial y} = \frac{D_h}{\mu_h} \frac{\partial \ln p(y)}{\partial y}. \tag{5.129}$$

This can now be integrated from the front surface to the back, as

$$V_y = -\int_0^L F_y \, dy = -\frac{k_B T}{e} \ln\left(\frac{p(L)}{p(0)}\right) = \frac{k_B T}{e} \ln\left(1 + \frac{\Delta p(0)}{p_0}\right). \tag{5.130}$$

This voltage is often referred to as the transverse photovoltage. If the excess density is small compared to the background hole concentration, the logarithm is easily expanded and the photovoltage is a direct measure of the excess density.

5.8. THE WORK FUNCTION

In Chapter 3, we first introduced the photoelectric effect, where incoming photons would cause electrons to be emitted from a material providing that the photon energy was sufficiently high. That is, the onset of current, due to the emission of electrons, is sensitive to the frequency of the light that is shone upon the surface. This led to the definition of the work function as the energy barrier over which electrons at the Fermi energy needed to be raised. On the other hand, the magnitude of the emission current depended on the intensity of the light — the number of photons per unit of time which reaced the surface. This is fine for a metal, but it is not obvious that the Fermi energy is the proper reference point in a semiconductor where the Fermi energy lies in a forbidden energy gap. Instead, we would assume that one of the band edges — either the conduction band or the valence band — would be the natural reference energy. In this section, we want to examine this in some more detail.

In semiconductors, as we have indicated, the photoemission process is a little more complicated due to the energy gap. This is indicated in Fig. 5.17, we show these energies. Naturally, the work function is the energy required to raise an electron from the Fermi energy to the vacuum level. The new energies are the *electron affinity* and the *ionization energy*. The electron affinity is the energy required to raise an electron from the bottom of the conduction band to the vacuum level, while the ionization energy is the energy required to

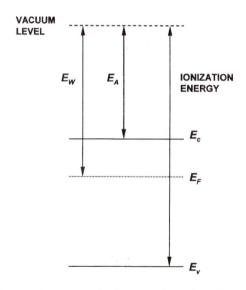

Figure 5.17. The important energy scales for a semiconductor that are relevant to any discussion of photoemission and the photoelectric effect. E_W is the *work function* of the semiconductor and is defined in a similar manner to that for metals. In contrast to a metal, however, we now have an additional two energy scales. The *electron affinity* (E_A) is the energy required to move an electron from the bottom of the conduction band to the vacuum level. The *ionization energy*, on the other hand, is the energy required to move an electron from the top of the valence band to the vacuum level.

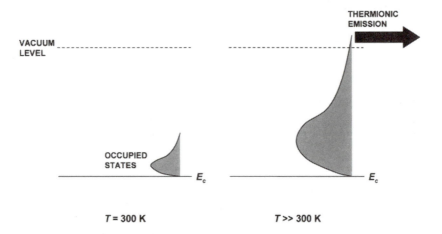

Figure 5.18. A schematic illustration of the distribution of occupied electron states (shaded region) in the conduction band of a semiconductor at room temperature. The form of this distribution results from *convolving* the Fermi function with the density of states in the conduction band. Thermionic emission corresponds to the thermal excitaton of electrons to the vacuum level.

raise an electron from the top of the valence band to the vacuum level. Clearly, the ionization energy is the electron affinity plus the band gap energy. Rather than discuss the photoelectric effect, we deal with thermionic emission, where carriers are thermally excited to the vacuum level. This is a high temperature process, but the important energies are the same as for photoemission. In Fig. 5.18, we schematically show a distribution function for the electrons in the conduction band (actually the product of the distribution function and the density of states, which is termed the particle distribution). The current that is emitted from the semiconductor is basically given by[9]

$$J_y = -e \int_{E_A}^{\infty} v_y \rho(E) f(E) \, dE. \tag{5.131}$$

The quantity E_A is the electron affinity, while $\rho(E)$ is the density of states. We take the polar angle relative to the y axis, so that $v_y = v \cos \theta$, so that the previous integration over θ, which led to the density of states now becomes

$$\int_0^{\pi/2} \cos\theta \sin\theta \, d\theta = \frac{1}{2}, \qquad (5.132)$$

instead of the factor of 2 that was previously obtained. The limits arise because we take only one direction of the velocity, not both (electrons moving away from the surface are not emitted). Thus, we may write (5.131) as

$$
\begin{aligned}
J_y &= -\frac{e}{4}\int_{E_a+E_c}^{\infty} v \frac{1}{2\pi^2}\left(\frac{2m^*}{\hbar^2}\right)^{3/2} (E-E_c)^{1/2} \exp\left(-\frac{E-E_F}{k_B T}\right) dE \\
&= -\frac{e}{2\pi^2}\frac{m^*}{\hbar^3}\int_{E_A+E_c}^{\infty} (E-E_c)\exp\left(-\frac{E-E_F}{k_B T}\right) dE.
\end{aligned} \qquad (5.133)
$$

In this last expression, we have assumed that the semiconductor is nondegenerate at the temperature of interest. This can be rewritten through the introduction of the reduced units $x = (E - E_c)/k_B T$, and $x_A = E_A/k_B T$. Then, (5.133) becomes

$$
\begin{aligned}
J_y &= -\frac{e}{2\pi^2}\frac{m^*}{\hbar^3}\exp\left(-\frac{E_c-E_F}{k_B T}\right)\int_{E_A+E_c}^{\infty}(E-E_c)\exp\left(-\frac{E-E_c}{k_B T}\right) dE \\
&= -\frac{e}{2\pi^2}\frac{m^*(k_B T)^2}{\hbar^3}\exp\left(-\frac{E_c-E_F}{k_B T}\right)\int_{x_A}^{\infty} x e^{-x}\, dx \\
&= -\frac{e}{2\pi^2}\frac{m^*(k_B T)^2}{\hbar^3}\exp\left(-\frac{E_c-E_F}{k_B T}\right)\left(1+\frac{E_A}{k_B T}\right)\exp\left(-\frac{E_A}{k_B T}\right) \\
&= -\frac{e}{2\pi^2}\frac{m^*(k_B T)^2}{\hbar^3}\left(1+\frac{E_A}{k_B T}\right)\exp\left(-\frac{E_W}{k_B T}\right).
\end{aligned} \qquad (5.134)
$$

Hence, even though it is the electron affinity that is important, it is the work function that arises in the final exponential weighting term. The coefficient can be written as AT^2, where

$$A = \frac{m^* e k_B^2}{2\pi^2 \hbar^3} = 119.8\left(\frac{m^*}{m_0}\right) A/(cm \cdot {}^\circ K)^2 \qquad (5.135)$$

is the Richardson–Dushman constant. The work function for Si is about 4.1 eV. If the surface is heated to 1000 K, then (5.134) predicts that the thermal emission current will be about 1.4×10^{-11} A/cm^2. This is a very small current density. Now, tungsten has a work function of 4.5 eV, which is larger than that of Si. However, there is no problem to heat this material to 2500 K. Then, the emission current is about 14 A/cm^2, which is some 12 orders of magnitude larger—a result solely from the ability to raise the temperature sufficiently high.

Before an electron is emitted from the semiconductor (or the metal, as in the case of tungsten), the solid is electrically neutral. After the emission of an electron, the surface is slightly positively charged due to the loss of the electron. This creates a Coulomb attraction that will try to pull the electron back to the surface. In general, however, this is not a big modification of the thermal emission theory already presented. It becomes more important when we try to use an electric field to pull the electrons out of the solid—field emission. In general, the electric field lowers the "vacuum level" away from the surface. However, the attractive Coulomb potential softens the potential into the gradual shape shown in Fig. 5.19. The attractive potential is, in fact, the attraction between the electron and its own *image charge*.

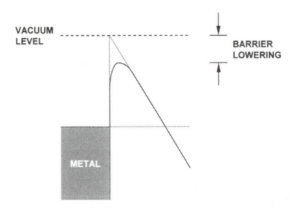

Figure 5.19. The surface potential that results from the *combined* effects of a repulsive applied electric field and the usual attractive force that the solid exerts on the electron. Note how the presence of the applied electric field has caused a *lowering* of the effective energy barrier for the emission of electrons from the solid.

The net of the interaction of the applied electric field and the image potential is to lower the effective barrier height, as shown in the figure. This process is knows as the *Schottky effect*. When the field is sufficiently high, the electrons can *tunnel* through this barrier — a process known as field emission. In general, the tunneling of the electrons can be written using only the exponential behavior from the tunneling coefficient. Hence, the tunneling coefficient is approximately

$$T \sim \exp(-2\gamma \, d_{\text{eff}}), \tag{5.136}$$

where γ is the decay constant of the evanescent wave function under the barrier, and d_{eff} is the effective thickness of the barrier at the Fermi energy (see Fig. 5.18).

In general, we can write the total potential as a sum of the work function, the Coulomb image potential, and the applied electric field, which is

$$V(z) = E_W - \frac{e^2}{16\pi\varepsilon_0 z} - eF_z z. \tag{5.137}$$

The factor of 16 occurs because the distance between the electron and its image is $2z$, which leads to an extra factor of 4 in the force. We can find the peak in the potential by setting the derivative of (5.137), with respect to distance, equal to zero. This leads to

$$\frac{dV}{dz} = 0 = \frac{e^2}{16\pi\varepsilon_0 z_{\text{max}}^2} - eF_z, \tag{5.138}$$

and

$$z_{\text{max}} = \sqrt{\frac{e}{16\pi\varepsilon_0 F_z}}. \tag{5.139}$$

This result can now be reinserted into (5.137) to give the maximum of the potential as

$$V_{\text{max}} = E_W - \sqrt{\frac{e^3 F_z}{16\pi\varepsilon_0}}. \tag{5.140}$$

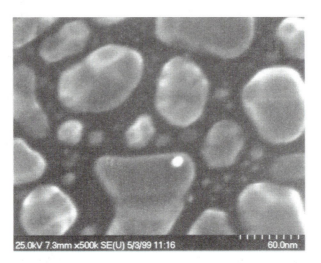

Figure 5.20. Field-emission scanning-electron microscope image of gold particles on a carbon film. It is clear from this picture, in which the spacer bar represents a distance of just 60 nm, that we can resolve the sharp edges of the gold particles with a resolution of roughly 1.5 nm. *Picture provided courtesy of S. Myhajlenko.*

It is clear that the electric field effectively lowers the barrier to the emission of electrons from the material (see Fig. 5.18).

The effective tunneling coefficient can be computed by evaluating the spatial variation of the decay constant for the evanescent wave, as

$$
\gamma d_{\text{eff}} = \int_0^{d_{\text{eff}}} \sqrt{\frac{2m[V(z) - E]}{\hbar^2}} \, dz
$$

$$
\approx \sqrt{\frac{2m}{\hbar^2}} \int_0^{E_W/eF_z} \sqrt{E_W - eF_z z} \, dt \qquad (5.141)
$$

$$
= \sqrt{\frac{2mV_{\text{max}}}{\hbar^2}} \frac{2E_W}{3eF_z}.
$$

Since the last factor in (5.141) is in the exponential factor for the tunneling coefficient, it is clear that the electric field plays the role of the temperature. Hence, we can write the field emission current as

$$J_{z,\mathrm{FE}} \sim AF_z^2 e^{-B/F_z}, \tag{5.142}$$

where B is a constant given by twice the value of the constants in (5.141) [twice because of the factor of 2 in the tunneling coefficient in (5.136)].

One of the most important uses of field emission is in the electron microscope. While most electron microscopes use thermal emitters to create the electron beam, high resolution instruments use field emitters. This is because the electrons come from the emission surface (the cathode) with a narrower range of energies in a field emitter than in a thermal emitter. In Fig. 5.20, we show an image of Au particles on a carbon film, in which the image was made with a field emission scanning electron microscope.

REFERENCES

1. C. Kittel, *Introduction to Solid State Physics*, 6th ed. (New York: Wiley, 1986).
2. D. K. Ferry, *Semiconductors* (New York: Macmillan, 1991).
3. J. S. Blakemore, *Semiconductor Statistics* (New York: Pergamon Press, 1962).
4. M. Abramowitz and I. A. Stegun, *Handbook of Mathematical Functions* (Washington, D.C.: National Bureau of Standards, AMS Series 55, 1964).
5. W. C. Dunlap, *An Introduction to Semiconductors* (New York: Wiley, 1957).
6. A. H. W. Beck and H. Ahmed, *An Introduction to Physical Electronics* (New York: Elsevier, 1968).
7. E. H. Hall, Am. J. Math. **2**, 287 (1879).
8. K. Seeger, *Semiconductor Physics*, 2nd ed. (Berlin: Springer-Verlag, 1982).
9. D. K. Ferry and R. Fannin, *Physical Electronics* (Reading, Mass.: Addison-Wesley, 1971).
10. K. von Klitzing, Phys. Rev. Lett. **45**, 494 (1980).

PROBLEMS

1. Solid copper is a good metal, which is finding increasing usage as an interconnect in modern integrated circuits. Copper contributes one electron per atom to the conduction band. What is the Fermi energy at $T = 0$? (Hint: you must first find the crystal structure, or find the number of atoms per unit volume using the molecular weight and Avogadro's number. You may assume the free electron mass.)

2. Germanium has a band gap of 0.7 eV, and, for pure (intrinsic) material, the Fermi level lies near the middle of the gap. What is the probability that a state in the conduction band is occupied at 0, 300, 500 K?

3. At temperatures just above 0 K, the Fermi–Dirac distribution can be expanded in a series about the the Fermi energy and integrated by parts. If one does this, show that for temperatures near (but not equal to) absolute zero, the Fermi energy is given by

$$E_F = E_{F0}\left[1 - \frac{\pi^2}{12}\left(\frac{k_B T}{E_{F0}}\right)^2\right].$$

4. In Bose–Einstein statistics, the Pauli exclusion principle does not hold. Therefore, there can be any number of particles in a given state. Obtain the number of ways in which N_i particles can be distributed among S_i states of energy E_i. Using this, and the fact that both the total number of particles and the total energy are constant, show that the Bose–Einstein distribution is

$$P_{\text{BE}} = \frac{1}{\exp(E/k_B T) - 1}.$$

5. What is the average energy of the electrons in the conduction band of a semiconductor? (Use nondegenerate statistics and compute the energy average as one would the number of carriers in the conduction band. Normalize the result to energy per electron.)

6. A Si crystal contains 10^{16} cm^{-3} phosphorous atoms, with an ionization energy of 10 meV. The energy gap is 1.1 eV. Determine the position of the Fermi energy and the number of free electrons in the conduction band at 300 and 500 K.

7. In a particular semiconductor, 37% of the donors are ionized. Is the Fermi level above the donor level or below it, and by how much?

8. For a Si electron mobility of 1500 cm^2/Vs, and a hole mobility of 500 cm^2/Vs, determine the collision times.

9. A piece of germanium is found to have a Hall constant of -7.5×10^{-3} m^3/C and a resistivity of 10 Ω-cm. Calculate the mobility, the density, and the type of the carriers.

10. A bar of *n*-type germanium is 1 cm long, and has a cross section of 1×1 mm^2. When it is placed in a magnetic field of 0.5 T, oriented along one of the short axes, a transverse field of 0.175 mV is measured when a current of 0.1 mA flows through the bar. The longitudinal voltage is 0.92 mV. What is the mobility of the electrons and what is the Hall constant? What is the carrier density?

11. Assume that you are given a value of the mobility ratio $b = \mu_e/\mu_h = 10$. Derive a relationship which yields the value of N_a at which the Hall constant is zero.

12. Plot the thermionic emission current for a Cu cathode and a BaO cathode as a function of temperature. Each cathode has an emitting surface of 1.5 cm^2.

13. What is the field strength required to reduce the surface barrier of a tungsten surface to an effective thickness of 15 nm? What is the tunneling probability at this field strength?

14. What are the low temperature wavelength ranges for photoemission from Cu, W, and Ta? In what portion of the spectrum do these lie?

CHAPTER 6

Semiconductor Devices

In this chapter, we will begin to study a variety of semiconductor devices. The theory of the operation of these devices is based on the electronic properties of Chapter 4 and the transport properties of Chapter 5. While these two chapters have provided the basic understanding of the properties of the semiconductor materials, it is the properties that we modify by building inhomogeneous materials that gives rise to the device properties. For example, we use the properties of the dopants—acceptors and donors—to build the most basic of devices, the $p-n$ junction diode. From this simple device, we build more complicated devices such as field-effect transistors and bipolar transistors (by bipolar we mean that two types of carriers of opposite charge are involved in the transport—of course, these are the electrons and the holes). A somewhat simple junction diode is made by merely placing a metallic layer on the semiconductor to produce the metal–semiconductor diode—the Schottky-barrier diode. This simpler version will also appear in field-effect transistors, particularly in the compound semiconductors, which do not have a useful native oxide.

We begin in the next section with the $p-n$ junction, discussing how it is formed, what the electrostatics do to create the desired behavior, and how current flows under applied bias potentials. We then use two such diodes to create the bipolar transistor, again studying the electrostatics and the manner in which current flows through the structure. After this, the Schottky-barrier diode is introduced, which is followed by the treatment of the metal–semiconductor field-effect transistor (FET). Once the field-effect transistor is understood, we are ready to use a metal-oxide–semiconductor (MOS) structure to control the current flow, and this enables the treatment of the MOSFET. From this, we return to the heterojunction FET in its high-electron-mobility formulation in GaAs–AlGaAs. The chapter is brought to an end by the discussion of the use of the MOSFET in simple memory and logic applications, illustrated by the methods in which these simple circuits come together naturally as part of the processing of the individual devices. No attempt is made at an exhaustive study of the circuits. Rather, only the simplest circuits are discussed in order to improve an understanding of how they arise from the device layout on the chip.

6.1. THE p–n JUNCTION

Suppose that two pieces of semiconductor material are brought together. One of these is n-type material, doped with donors such as phosphorus or arsenic. The second material is p-type material, doped, for example, with boron. In isolation, the Fermi level in the n-type material is quite close to the conduction band, whereas that in the p-type material is quite close to the valence band. The immediate question we need to answer is: How do the bands align when these two materials are joined together? The answer to this question lies in the fact that, in thermal equilibrium, the Fermi energy must be constant in position throughout the adjoined material. We can readily show this by recalling the equation for the flow of holes in (5.116), which we can rewrite as (recall that we use F for the electric field, to avoid confusion with the energy E)

$$J_h = pe\mu_h F - eD_h \frac{\partial p}{\partial x}. \qquad (6.1)$$

In general, the hole density is given by the Fermi–Dirac statistics, but with the Fermi energy representing some local variation with position. In analogy with (5.39), we may write this density as

$$p = N_v \exp\left(\frac{E_v - E_F}{k_B T}\right) = n_i \exp\left(-\frac{\Delta}{k_B T}\right), \qquad (6.2)$$

where $\Delta = E_F - E_{Fi}$, and the latter is the intrinsic Fermi energy. Using this in (6.1), and inserting the Einstein relation for the holes, leads to

$$J_h = pe\frac{D_h e}{k_B T}\frac{1}{e}\frac{\partial E_{Fi}}{\partial x} + eD_h \frac{p}{k_B T}\frac{\partial(E_F - E_{Fi})}{\partial x} = \frac{peD_h}{k_B T}\frac{\partial E_F}{\partial x}. \qquad (6.3)$$

Here, it is important to note that the electric field F arises from a local potential that is defined by the variation of the *intrinsic* Fermi energy. In electrical equilibrium, where there is no applied voltage, there is no current flow, and we require

$$\frac{\partial E_F}{\partial x} = 0. \qquad (6.4)$$

Hence, the Fermi energy must be constant throughout the p–n junction, in the absence of any current flow (no applied voltages). We show this in Fig. 6.1.

Although one cannot easily adjoin two dissimilar pieces of semiconductor, as predicated above, it is relatively easy to vary the doping concentrations of donors and acceptors within a single piece of semiconductor (we will illustrate this at the end of this section). It can be seen from Fig. 6.1(b) that there is significant band bending in the junction, and the variation of E_{Fi} from one side of the junction to the other leads to a large *built-in* potential. This potential also causes the variations in the conduction and valence band from one side of the junction to the other. When the junction is formed, there is a large population of holes on one side of the junction (the p-type side), and a small minority concentration on the other side. Holes will diffuse from the large population region to the small population region. This leaves excess negatively charged acceptors on the p-type side of the junction and excess positively charged holes on the other

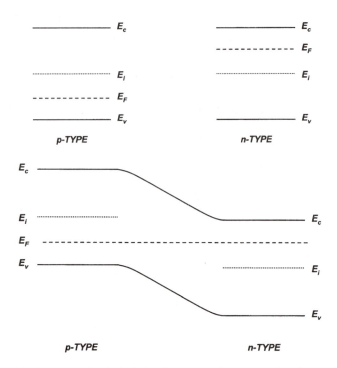

Figure 6.1. (a) The energy bands in isolated n-type and p-type semiconductors. Note the position of the Fermi level in these samples, relative to its intrinsic value. (b) The energy bands in a p–n junction at thermal equilibrium. The requirement that the Fermi level be *constant* throughout the junction leads to position-dependent *bending* of the conduction and valence bands.

side. By the same token, there is a large concentration of electrons on the n-type side of the junction, and a small minority population on the other side. Consequently, electrons will flow from the large density region to the low density region, leaving excess positively charged donors on the n-type side and excess negatively charged electrons on the other side.

The net result of this particle diffusion across the junction is that there is extra positively charged species in the n-type region and excess negatively charged species in the p-type region. This creates a *space charge* in the junction region. In general, the net charge in any semiconductor is given by

$$\rho = e(N_{Di} + p - N_{Ai} - n), \tag{6.5}$$

where the subscript i stands for "ionized." Normally, in the n-type semiconductor, the ionized donors are compensated by the electrons. In the p-type semiconductor, the ionized acceptors are compensated by the holes. Near the junction, however, the electrons have flowed out of the n-type region into the p-type region, and the holes have flowed the reverse direction. This diffusion across the junction causes the space-charge non-neutrality, which leads to a built-in potential according to Poisson's equation

$$\frac{d^2V(x)}{dx^2} = -\frac{\rho(x)}{\varepsilon},$$ (6.6)

where ε is the dielectric permittivity. This built-in potential varies with position, so that it leads to an electric field that *opposes* further diffusion of the carriers. The carriers diffuse just sufficiently to provide exactly the right amount of built-in potential that the Fermi energy can be constant with position. This small detailed balance is a result of the importance of the statistics developed in the last chapter.

In fact, we can usually consider that the electrons that leave the n-type region will recombine with holes in the p-type region. By the same token, holes that leave the p-type region will recombine with electrons in the n-type region. As a result, the space charge is primarily provided by the unneutralized donors and acceptors in the region adjacent to the junction. This region is called the *space-charge region*, and exists in the region where the conduction and valence bands have their bending in Fig. 6.1(b). The size of the built-in potential is then given by the separation of the conduction band (or the valence band) on the two sides of the junction, or

$$eV_{bi} = E_{c,p} - E_{c,n} = E_{v,p} - E_{v,n}.$$ (6.7)

Before proceeding to the electrostatics of the junction, let us first consider how a junction can be made. The earliest $p–n$ junctions were made by introducing dopants during the bulk growth of the semiconductor itself. While an easy approach, it is not useful in today's fabrication of millions of diodes and transistors on a single chip. Thus, we consider the "diffused," or "implanted" diode structure.

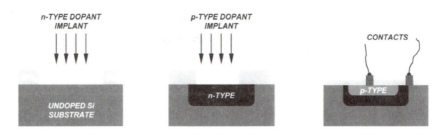

Figure 6.2. Schematic illustration showing the key fabrication steps that are used to realize an implanted planar p–n junction. Details of the fabrication flow are described in the text.

Here, we discuss a diode made in nominally undoped Si, with the p-type region at the surface. The process steps are shown in Fig. 6.2. First, an oxide layer is grown on the Si. This layer will serve to define the lateral extent of the n-type region. A hole is opened in the oxide by a lithography process, in which a photo-resist is deposited and exposed so that the oxide can be chemically removed just in the area where the hole is desired, as shown in the left panel of Fig. 6.2. Then, donor atoms are either diffused into the material in a hot oven, or they are "thrown into" the semiconductor after being accelerated to high voltage—the ion implantation process. In either case, a distribution of donors is introduced into the semiconductor. Then, the first oxide is removed, and a second oxide is grown. This is then subjected to a second lithography step to open a hole used to define the p-type region. This region is now defined again by either diffusion or implantation, as indicated in the central panel of Fig. 6.2. However, it is important to have the number of acceptors larger than that of the donors, so that there is a net excess of acceptors near the surface, a point to which we return below. Finally, the second oxide is removed and a third oxide formed. This is again patterned to provide the locations of the contacts to the two regions, as shown in the right-hand panel of Fig. 6.2. More details on the diffusion and ion implantation processes are given in Appendix B.

In Fig. 6.3(a), the completed device has been turned on its side. The doping concentrations of acceptors and donors are shown in Fig. 6.3(b). First, the donor concentration must be more than the background doping to define the n-type region. As noted earlier, the number of acceptors must be still larger, at least near the surface, so

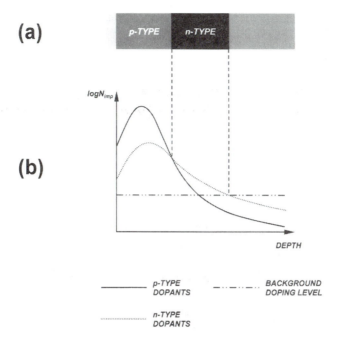

(a)

(b)

p-TYPE   _n-TYPE_

$logN_{imp}$

DEPTH

————	_p-TYPE_ _DOPANTS_	_— ·· — ·· —_ _BACKGROUND_ _DOPING LEVEL_
··············	_n-TYPE_ _DOPANTS_	

Figure 6.3. Spatial variation of the doping profile in an implanted p–n junction. At any given depth from the surface, the type of carriers (electrons or holes) that dominate transport is determined by the relative concentration of n- and p-type dopants and the background doping.

that the latter region is p-type. Both the two doping concentrations, and the physical dimensions of the junction, can be controlled by the processing time and temperature. For example, in the diffusion process, the temperature and environment determine the depth and concentration, respectively. In the implanation process, these are controlled by the accelerating voltage and charge dose (product of beam current and exposure time), respectively.

6.1.1. Electrostatics of the p–n Junction

The built-in potential described in the preceding appears because the junction shifts the relative positions of the energy levels of the two regions on either side of the junction. As a result, electrons in the p-region are at a higher energy than those in the n-region. Consequently, there are fewer of them in the former region. The junction potential is given by this shift of the energy levels. For example, we can write the electron density in the n-region as

$$n_n = N_c \exp\left(-\frac{E_{cn} - E_F}{k_B T}\right) = N_c \exp\left(-\frac{E_{cp} - eV_{bi} - E_F}{k_B T}\right). \quad (6.8)$$

At the same time, the electron density in the p-region is given as

$$n_p = N_c \exp\left(-\frac{E_{cp} - E_F}{k_B T}\right) = n_n \exp\left(-\frac{eV_{bi}}{k_B T}\right). \quad (6.9)$$

Here, we have used (6.8) to reach the last expression. As a result, we can relate the densities on either side of the junction to the built-in potential through

$$\frac{n_p}{n_n} = \frac{p_n}{p_p} = \exp\left(-\frac{eV_{bi}}{k_B T}\right). \quad (6.10)$$

Now, we can use the above relations to write the built-in potential as

$$\begin{aligned} eV_{bi} &= E_{cp} - E_{cn} = E_G + E_{vp} - E_{cn} \\ &= E_G + (E_{vp} - E_F) - (E_{cn} - E_F) \quad (6.11) \\ &= E_G + k_B T \ln\left(\frac{N_A}{N_v}\right) + k_B T \ln\left(\frac{N_D}{N_c}\right), \end{aligned}$$

where we have introduced (5.55) and its equivalent for p-type material. The last two terms can be combined, so that (normally, N_D, $N_A < N_c$, N_v)

$$eV_{bi} = E_G - k_B T \ln\left(\frac{N_v N_c}{N_A N_D}\right) = k_B T \ln\left(\frac{N_A N_D}{n_i^2}\right). \quad (6.12)$$

In the last line, we have inserted the relationship (5.43) between the intrinsic concentration and the energy gap. Thus, the built-in potential is determined by the doping levels on the two sides of the junction. Consider a Si junction, in which the n-type region is doped

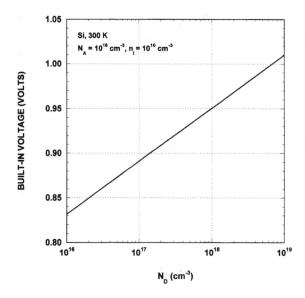

Figure 6.4. Computed variation of the built-in voltage with doping density in a silicon *p–n* junction at 300 K. In this plot, the acceptor concentration N_A is assumed to be 10^{18} cm^{-3} and the donor density N_D is varied.

to 10^{18} cm^{-3}, and the *p*-type region is doped to 10^{17} cm^{-3}. From (6.12), the built-in potential will be about 0.87 V (assuming the band gap is 1.08 eV), which is less than the band gap at room temperature (using the values for the effective densities of states from the last chapter). In Fig. 6.4, we plot the built-in voltage as a function of donor doping, for a fixed value of acceptor doping.

With this value as a boundary condition, we can now solve Poisson's equation for the actual potential variation through the junction region. However, if we are to proceed, we need to make an approximation for the charge density. While the exact spatially varying charge density can be used to solve this equation, this normally requires computational tools. Here, we make a simple approximation that allows a straight-forward analytical result to be obtained. This approximation is shown in Fig. 6.5, where we plot the charge density. On the *n*-type side of the junction, we *assume* that all the electrons are removed for a distance x_n, leaving only positively charged donors in this region. Similarly, we assumed that all the

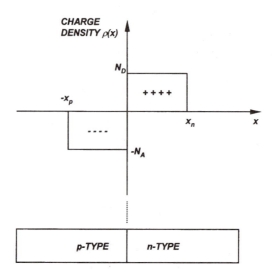

Figure 6.5. The charge density distribution for the *p–n* junction in the *abrupt junction* approximation. Diffusion of holes into the *n*-type region leaves the *p*-type side of the depletion region negatively charged while the diffusion of electrons into the *p*-type region leaves the *n*-type side of the depletion region positively charged.

holes have been removed for a distance x_p on that side of the junction. The two quantities, x_n and x_p, represent the *depletion depths* on each of the two sides of the junction. The total *depletion width* of the junction will be the sum of these two depths. We can write the space-charge density as

$$\rho(x) = \begin{cases} eN_D, & 0 < x < x_n, \\ -eN_A, & -x_p < x < 0. \end{cases} \tag{6.13}$$

The space-charge density is assumed to be zero outside this region near the junction. As a result, we can integrate Poisson's equation once, to give

$$F_x = -\frac{dV}{dx} = \begin{cases} \dfrac{eN_D}{\varepsilon}(x + C), & 0 < x < x_n, \\ -\dfrac{eN_A}{\varepsilon}(x + D), & -x_p < x < 0. \end{cases} \tag{6.14}$$

The quantities C and D are constants of the integration, which we need to evaluate. At the two ends of the depletion region (x_n and $-x_p$), the electric field must vanish. Thus, we find that

$$C = -x_n, \qquad D = x_p \tag{6.15}$$

are required to satisfy these two boundary conditions. This leads to another consequence that is vitally important. The electric field must be continuous at $x = 0$, and the two parts of (6.14), together with (6.15), lead to

$$N_D x_n = N_A x_p. \tag{6.16}$$

In fact, this simple statement also tells us *that the total depletion region is space charge neutral* in the aggregate. That is, the total amount of positive charge on the *n*-type side of the junction must equal the total amount of negative charge on the *p*-type side of the junction. If this were not the case, the overall semiconductor would be charged and move around in stray fields. This just doesn't occur, and global charge neutrality is required for any semiconductor device. As a consequence, the peak value of the electric field occurs at $x = 0$, and this value is given by

$$F_{\text{max}} = F(0) = -\frac{e N_D x_n}{\varepsilon} = -\frac{e N_A x_p}{\varepsilon}. \tag{6.17}$$

The electric field is plotted in Fig. 6.6(a) showing how this quantity varies across the junction.

We can now integrate (6.14) to obtain the potential itself across the junction. Again, there are two integration constants, and the result is

$$V(x) = \begin{cases} -\dfrac{e N_D}{2\varepsilon}(x - x_n)^2 + C_1, & 0 < x < x_n, \\[2mm] \dfrac{e N_A}{2\varepsilon}(x + x_p)^2 + D_1, & -x_p < x < 0. \end{cases} \tag{6.18}$$

Here, we make a choice for the reference level for the potential. To set this, we use the *p*-type region, and assume that $V(x) = 0$ at

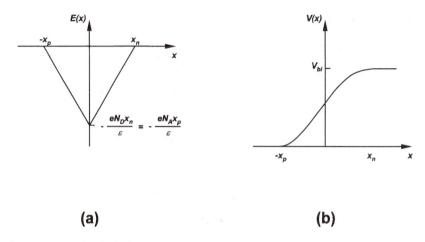

(a)   **(b)**

Figure 6.6. (a) Electric field variation in the p–n junction in the abrupt-junction approximation. Since the regions away from the depletion region are assumed to be neutral, the Poisson equation dictates that the electric field in these regions must be equal to zero. (b) Corresponding variation of potential in the p–n junction.

$x = -x_p$. Consequently, $D_1 = 0$. At the other end of the depletion region, the voltage is just the built-in potential, so that

$$C_1 = V_{bi}. \tag{6.19}$$

In Fig. 6.6(b), we plot the spatial variation of the potential.

We can determine another property by using the fact that the potential must be continuous at $x = 0$, or

$$\frac{eN_A x_p^2}{2\varepsilon} = -\frac{eN_D x_n^2}{2\varepsilon} + V_{bi}. \tag{6.20}$$

This can be rearranged to give

$$V_{bi} = \frac{eN_A x_p^2}{2\varepsilon} + \frac{eN_D x_n^2}{2\varepsilon}. \tag{6.21}$$

The total width of the junction depletion region is $W = x_n + x_p$. This last equation can be rearranged to determine this width, which will depend just upon the built-in potential and the doping concentrations. To begin, we rewrite (6.21) as

$$\frac{2\varepsilon V_{bi}}{e} = N_A x_p^2 + N_D x_n^2 = N_A x_p^2 + N_D \left(\frac{N_A}{N_D} x_p\right)^2$$

$$= \frac{N_A(N_A + N_D)}{N_D} x_p^2 = \frac{N_D(N_D + N_A)}{N_A} x_n^2. \tag{6.22}$$

Now, we can write the two depletion depths as

$$x_p = \sqrt{\frac{2\varepsilon V_{bi}}{e} \frac{N_D}{N_A(N_D + N_A)}},$$

$$x_n = \sqrt{\frac{2\varepsilon V_{bi}}{e} \frac{N_A}{N_D(N_D + N_A)}}. \tag{6.23}$$

Finally, we can write the total width of the depletion region as

$$W = x_n + x_p = \sqrt{\frac{2\varepsilon V_{bi}}{e(N_D + N_A)}} \left[\sqrt{\frac{N_D}{N_A}} + \sqrt{\frac{N_A}{N_D}}\right]$$

$$= \sqrt{\frac{2\varepsilon V_{bi}}{e(N_D + N_A)}} \left[\frac{N_D + N_A}{\sqrt{N_D N_A}}\right] \tag{6.24}$$

$$= \sqrt{\frac{2\varepsilon V_{bi}}{e} \left(\frac{N_D + N_A}{N_D N_A}\right)}.$$

This can be rearranged to give the built-in potential in terms of the junction width W as

$$V_{bi} = \frac{eW^2}{2\varepsilon} \left(\frac{N_D N_A}{N_D + N_A}\right). \tag{6.25}$$

For the values of the doping given above (10^{18} cm^{-3} in the n-type region and 10^{17} cm^{-3} in the p-type region), we can determine the values for the parameters as $x_n = 10.2$ nm, $x_p = 102.5$ nm, and $W = 112.7$ nm. If the doping was lowered to 10^{17} cm^{-3} in both regions, then V_{bi} would be only 0.81 V, and the two depletion widths would be increased to 73.3 nm, and $W = 0.147$ μm. An important point from (6.24) is that the *lower-doped region* produces the domi-nant part of the junction width W. The junction width W is plotted

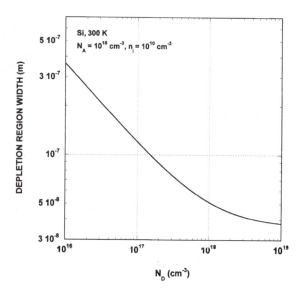

Figure 6.7. Computed variation of the depletion-region width with doping density in a silicon p–n junction at 300 K. In this plot, the acceptor concentration N_A is assumed to be 10^{18} cm^{-3} and the donor density N_D is varied.

in Fig. 6.7 for a range of donor doping levels, for a fixed acceptor doping level (the same levels are used as for Fig. 6.4).

6.1.2. Current Flow in *p–n* Junctions

The direction of the built-in potential of the junction is such that it is positive on the *n*-type side of the junction. Thus, if we apply an external potential with the positive terminal connected to the *p*-type region, the overall potential will be *reduced*. The reduction in the potential will cause more electrons to move across the junction from the *n*-type region to the *p*-type region, and more holes from the *p*-type region to the *n*-type region. That is, the applied potential lowers the built-in potential, the latter of which was built up sufficiently to oppose such motion by the majority carriers. We term this correction of the voltage *forward bias*. The forward bias results in very large currents, as we shall shortly see. On the other hand, if we apply the external potential with the positive terminal connected to the n-type region, the result is an increased potential across the junction. This *reverse* bias also causes a current flow, but only a small one. As a result, the current that results from these two *directions* of bias are quite different in order of magnitude. One direction of bias produces large currents, while the opposite polarity produces only small currents. This lack of symmetry in the current response is known as *rectification*. That is, high conductance and large currents result from only one direction of the bias, not from both. The definition of forward bias is that which provides the large current flow in response to the applied voltage.

We can understand how the current responds to the voltage by considering (6.10), which describes the ratio of the carrier densities on the two sides of the junction. To (6.10), we will add a subscript to the minority carrier concentration. Hence, we rewrite (6.10) as

$$\frac{n_{p0}}{n_n} = \frac{p_{n0}}{p_p} = \exp\left(-\frac{eV_{bi}}{k_B T}\right). \tag{6.26}$$

That is, the built-in voltage defines this ratio in the equilibrium case, where there is no applied voltage and no current flow. In the presence of an applied voltage, however, this equation is modified, and becomes

$$\frac{n_p}{n_n} = \frac{p_n}{p_p} = \exp\left(-\frac{e(V_{bi} - V_a)}{k_B T}\right), \tag{6.27}$$

where the sign of the applied voltage V_a has been taken to oppose the built-in potential as already discussed. Thus, we can write the actual minority carrier density in the two regions as

$$\Delta n = n_p - n_{p0} = n_n \left[\exp\left(-\frac{e(V_{bi} - V_a)}{k_B T}\right) - \exp\left(-\frac{eV_{bi}}{k_B T}\right)\right]$$

$$= n_n \exp\left(-\frac{eV_{bi}}{k_B T}\right)\left[\exp\left(\frac{eV_a}{k_B T}\right) - 1\right] \tag{6.28}$$

$$= n_{p0}\left[\exp\left(\frac{eV_a}{k_B T}\right) - 1\right]$$

and

$$\Delta p = p_{n0} \left[\exp\left(\frac{eV_a}{k_B T}\right) - 1 \right]. \tag{6.29}$$

The quantities Δn and Δp are the *injected* concentrations that result from the applied bias. These injected carriers are referred to as *excess* carriers, since they represent a change of the local minority concentrations away from their equilibrium values.

In Sec. 5.6, the motion of the excess carriers was discussed in terms of carrier diffusion. Indeed, from that treatment, we can say that the excess electrons, injected into the p-type region, diffuse toward the negative x direction. This produces a current in the positive x direction, in agreement with the direction assumed for the potential. The electron diffusion current is given by

$$J_e = eD_e \frac{\partial \Delta n}{\partial x} = eD_e \frac{\partial}{\partial x}(\Delta n e^{(x+x_p)/L_h}), \qquad x < -x_p. \tag{6.30}$$

Here, we have introduced the corresponding electron diffusion length from (5.126). We note that the injection is assumed to occur at the left edge of the depletion region; that is, at the point where the p-region adjoins the depletion region between the two sides of the junction. While there is no real justification to support this treatment, we have assumed that there are no free carriers within the depletion region, so the extra electrons have to appear on the p-region side of this depletion region. The lowered value of the net potential has, however, narrowed the depletion region, as can be seen from (6.24). The number of excess carriers decreases as one moves away from the junction, according to the exponential decay shown in (6.30). This would imply a nonuniform current, if there were no other current flow. However, the decrease in Δn arises from recombination, so there must be a majority carrier (drift) current of holes to provide the carriers necessary for this recombination. Hence, the spatially constant current density is set by the injected carrier density at the edge of the depletion region, and

$$J_e = \lim_{x \to -x_p} eD_e \frac{\partial}{\partial x}(\Delta n e^{(x+x_p)/L_h}) = \frac{eD_e n_{p0}}{L_n} \left[\exp\left(\frac{eV_a}{k_B T}\right) - 1 \right]. \tag{6.31}$$

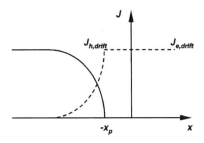

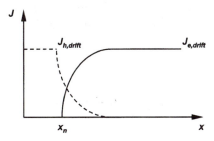

Figure 6.8. Decay of the injection (diffusion) carriers and their current. (a) Injected holes recombine with drifting electrons. (b) Injected electrons recombine with drifting holes.

This is shown in the top panel of Fig. 6.8.

A similar current flows for the holes injected into the n-region. Here, we use x_n and L_p, but the equivalent equation to (6.31) is

$$J_e = - \lim_{x \to x_n} eD_h \frac{\partial}{\partial x} (\Delta p e^{-(x-x_n)/L_p}) = \frac{eD_h p_{n0}}{L_p} \left[\exp\left(\frac{eV_a}{k_B T} \right) - 1 \right]. \quad (6.32)$$

These two currents and the corresponding hole diffusion current and the electron drift current that produces recombination carriers are shown in the bottom panel of Fig. 6.8. and the total current for each particle type is shown in Fig. 6.9. The two preceding equations serve to give us the total current density as

$$J = \left(\frac{eD_h p_{n0}}{L_p} + \frac{eD_e n_{p0}}{L_n} \right) \left[\exp\left(\frac{eV_a}{k_B T} \right) - 1 \right]. \quad (6.33)$$

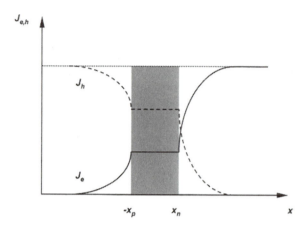

Figure 6.9. The variation with position of the *total* electron and hole currents in a forward-biased p–n junction. The shaded region corresponds to the depletion region.

It is important to note that while both drift and diffusion of electrons and holes occur in the junction, the current level in (6.33) is produced *solely* by the diffusion of the injected carriers. The drift currents exist solely to provide the needed carriers to recombine with these injected carriers.

It is the exponential term in the square brackets that produces the rectification in the diode. We plot the total current density in Fig. 6.10. When the device is forward biased, the current is largely determined by the size of this exponential term. On the other hand, when the device is reversed biased, the current saturates at a value termed the reverse saturation current

$$I_{\text{sat}} = -A\left(\frac{eD_h p_{n0}}{L_p} + \frac{eD_e n_{p0}}{L_n}\right), \qquad (6.34)$$

where A is the cross-sectional area of the current carrying region of the diode. In the preceding discussion, we assumed that the n-type side was doped to 10^{18} cm^{-3}, and the p-type side was doped to 10^{17} cm^{-3}. From Appendix C, we know that $n_i = 1.0 \times 10^{10}$ cm^{-3} at 300 K. Thus, $n_{p0} = 2.37 \times 10^3$ cm^{-3}, and $p_{n0} = 237$ cm^{-3}. It is clear that there are not many minority carriers present in thermal equilibrium. At these doping densities, the mobility of electrons and holes is

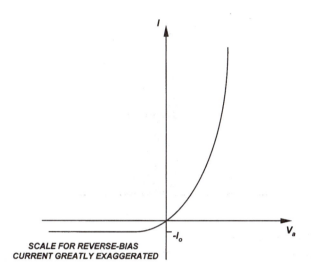

Figure 6.10. Ideal rectifying current–voltage characteristic for a *p–n* junction. Note that the current scale for the reverse-biasing situation is greatly exaggerated; unless breakdown occurs, the reverse-bias current is typically many orders of magnitude smaller than the forward-bias current.

about 250 and 100 cm²/Vs, respectively. This leads to diffusion constants of $D_e = 6.5$ cm²/s and $D_h = 2.6$ cm²/s. If the recombination time is 1 μs, then $L_e = 25$ μm and $L_h = 16$ μm. Now, this leads to a saturation current density of about 16 nA/cm². If the area of the device is 1.0 μm², then the saturation current is 1.6×10^{-16} A. This is an incredibly small current, and various leakage mechanisms would cause a significant increase in this value. But, these are beyond our current treatment. Now, while the reverse current is small, a forward bias dramatically changes this. For 1 V of forward bias, the forward current is 9.4 A! That is, 1 V of forward bias has increased the current by almost 17 orders of magnitude! This is truly rectification.

6.1.3. Diodes under Large Reverse Bias

In the reverse bias direction, where the applied bias serves to *increase* the potential across the depletion region, we have argued that the current flow in the *p–n* junction is almost negligible. In the preceding example where both sides of the junction were heavily doped, we found a reverse saturation current of only about 160 attoampere (the prefix *atto* means 10^{-18}, which should not be confused with *pico* and *femto*, which refer to 10^{-12} and 10^{-15}, respectively). However, when the electric field is sufficiently large, impact ionization and avalanche breakdown can occur. *Impact ionization* is a process by which a carrier is accelerated by an electric field to an energy sufficiently large that it can scatter with a valence electron and transfer an energy $> E_{gap}$ to the latter electron. This "kicks" the valence electron into the conduction band, which creates an additional conduction electron and a valence hole, both of which can now contribute to the current. Needless to say, the initial electron must gain an energy (from the electric field) larger than E_{gap} if it is to give up this amount to the valence electron, and this requires very high electric fields. If each of the new carriers, as well as the initial electron, are all then accelerated in the field, they can again undergo impact ionization. These three carriers then create six new carriers, and this process leads to exponential growth of the number of free carriers. This exponential growth is termed *avalanche breakdown*, since an enormous current will begin to flow in the reverse direction.

In Fig. 6.11(a), we plot the current flow through a p–n junction. When the reverse bias exceeds V_{bd}, the avalanche current begins to flow and this creates a large reverse current in the device. We can estimate when this will occur. Impact ionization begins to occur when the field exceeds about 200–300 kV/cm. For the heavily-doped diode parameters above, the latter field is more normal. In the case with no applied bias, we can evaluate the electric field from (6.17). Using the parameters for the unequally doped junction, we find a peak electric field of 154 kV/cm, with no bias applied! If we now apply a reverse bias, as shown in Fig. 6.11(b), to increase the electric field to 300 kV/cm, we find that the depletion width boundaries are similarly increased. The new values are $x'_n = 20$ nm, $x'_p = 200$ nm, and $W = 220$ nmn. Using (6.24) with $V_{bi} \rightarrow V_{bi} - V_a$, remembering that $V_a < 0$, we find that $V_{bi} - V_a = 3.31$ V. Hence, taking out the built-in potential of 0.87 V, we find that breakdown occurs for a reverse bias of $V_a = -2.44$ V. In Fig. 6.12, we plot the breakdown voltage for junctions with strongly asymmetric doping as a function of the lightly doped layer impurity concentration.

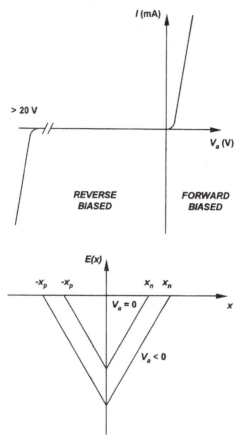

Figure 6.11. (a) Form of the current voltage characteristic for a realistic *p–n* junction, showing the breakdown that occurs at high reverse bias. (b) A reverse bias voltage adds to the built-in voltage and so increases the magnitude of the electric field in the depletion region. This effect is illustrated here, where we see that forward bias also increases the *width* of the depletion region.

Avalanche breakdown is generally found in lightly doped to moderately doped diodes. In very heavily doped diodes, where the total depletion width is only a few nanometers or less, another process can occur. This is Zener tunneling,[1] in which a valence electron in the valence band can tunnel through the band gap to the conduction band. Hence for this to occur, we generally need a sufficiently high reverse bias that the conduction band on the *n*-type side lies below the top of the valence band on the *p*-type side.

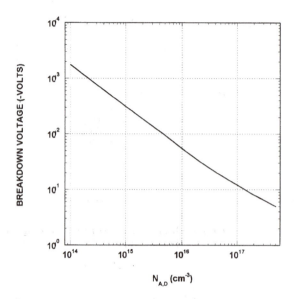

Figure 6.12. Typical variation of the breakdown voltage with doping level in a p–n junction with strongly asymmetric doping.

However, this is not sufficient, as it can always occur under reverse bias. More importantly, the physical distance between the conduction band on the n-type side and the valence band on the p-type side must be sufficiently small for tunneling to occur. This process is sketched in Fig. 6.13. In the preceding example, the width of the junction at breakdown was 220 nm. This is clearly too large a distance through which to tunnel. If, on the other hand, this depletion width were only about 5 nm, or less, then tunneling could occur. We can estimate the doping required for this by considering that the field will be larger than $E_{gap}/eW > 2 \times 10^6$ V/cm. For equally doped regions, this requires $N_D > 5 \times 10^{19}$ cm^3, which is approaching a significant fraction of the Si atoms in the crystal. Nevertheless, diodes have been made in which Zener tunneling is observed to occur. One might think, however, that these very high fields would also initiate impact ionization. This does not occur, however, because the impact ionization process has a mean free path that is some tens of nanometers or more. In these very thin depletion regions, the carrier tunnels before it can impact ionize, and is out of the high-field region once it tunnels.

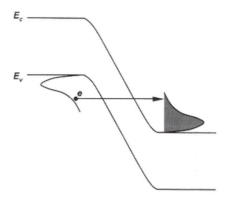

Figure 6.13. The Zener breakdown mechanisms is an *inter*band tunneling process in which carriers tunnel through the depletion region to occupy available energy states.

6.2. THE BIPOLAR TRANSISTOR

The bipolar junction transistor (BJT) is composed of two back-to-back $p-n$ junctions, which create a $p-n-p$ or $n-p-n$ structure. The former was the earliest one made, but the latter began to dominate in later years due to its improved performance, since electrons, rather than holes, provide transport within the device. We show a schematic of the $n-p-n$ BJT in Fig. 6.14. It was the bipolar transistor, made with metal point contacts for the two outer regions, that was the first transistor invented by Bardeen, Brattain and Shockley[2,3]. But, why should we conceive of such a structure as being useful? Consider, for example, Eq. (6.33). The reverse current through the $p-n$ junction depends linearly on the minority carrier densities in each of the two regions. These quantities are typically only a few thousand carriers per cubic centimeter, as we found in the earlier examples. However, if we could controllably vary these quantities, we could controllably vary the (reverse) current through the $p-n$ junction. This is the goal of the bipolar structure of the transistor, shown in Fig. 6.14. The second junction, the $p-n$ junction in a $n-p-n$ structure, is reverse biased, so that it is operating in the saturated-current regime. Now, we found that a forward-biased junction can increase the minority carrier concentration by many orders of magnitude, and this can be done in a reasonably controlled fashion through the forward-bias voltage. That is, if we forward bias the first junction, the $n-p$ junction, then we dramatically increase n_p on the right-hand side of this junction (in the central, or *base*, region). (The left-hand n-region in Fig. 6.14 is termed the *emitter*.) These excess carriers diffuse away from the junction, and if the second $p-n$ junction is placed closer than the diffusion length L_e, then we can increase the minority carriers on the left-hand side of the second junction (we term the right-hand n-region in Figs. 6.14 and 6.15 the *collector*). This, in turn, increases the reverse current through this second junction, as desired. This current results from the fact that the large reverse bias sucks these electrons across the junction into the output n-region, as shown in Fig. 6.15. As a result, we modulate the output current with the voltage across the input $n-p$ junction. The key factor we desire is that most of the input current through the $n-p$ junction will appear as output current in the $p-n$ junction. Only the small difference between these two currents will flow out through the connection to the central p-region. If we actually use this p-region current (the base current) as the input current, then we produce a large current gain between this

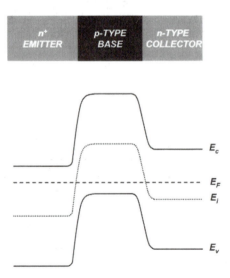

Figure 6.14. Shown top is a schematic diagram of a $n-p-n$ bipolar junction transistor. Note that in reality the base region is very much thinner than the emitter. At the bottom we show the energy-band diagram for the junction transistor at thermal equilibrium (note that, as required, the Fermi level is constant throughout the transistor).

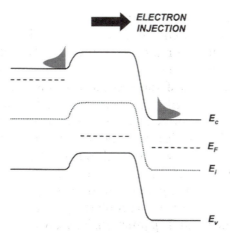

Figure 6.15. With the emitter-base junction forward biased and the collector-base junction reverse biased, electrons are injected from the emitter, travel through the base, and arrive at the collector.

current and the output current (the *collector* current). The most common set of output characteristics, however, is that for voltage bias on the input n–p junction, and an example of this is shown in Fig. 6.16.

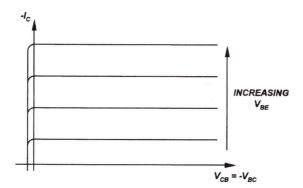

Figure 6.16. Schematic illustration showing the current–voltage characteristics of a n–p–n bipolar junction transistor at a number of different emitter-base voltages. Note that *negative* current is actually plotted on the vertical scale. Note how, as the emitter-base voltage is increased, the efficiency of electron injection from the emitter increases.

As pointed out, the central region, the *p*-type region in the present discussion, is termed the *base*. This name seems to have a historical basis as the earliest point-contact transistors had the other two contacts pressed against this central region, so it was called the base (upon which the rest depended). The input *n*-region is termed the *emitter*, as it is the source of the carriers upon which the operation of the transistor depends. In an equally obvious convention, the output *n*-region is termed the *collector*. These terminologies are applied whether the transistor is an *n–p–n* or a *p–n–p* device.

6.2.1. Current Flow in the BJT

In Fig. 6.17, we show the bias voltages and currents that result from these biases. Note, however, that the convention of voltage and current for the collector junction are to be taken very carefully. While the voltage V_{CB} will, in fact, be positive, it is a reverse bias of the base–collector *p–n* junction. Similarly, while the direction indicated for I_C is appropriate for forward bias of this junction, this current will actually be negative. We define a current for each junction through (6.33) and the appropriate area of the junctions. For the indicated direction of I_C, we may write down the current as

$$I_C = I_{C0}\left[\exp\left(-\frac{e|V_{CB}|}{k_B T}\right) - 1\right] - \alpha I_{E0}\left[\exp\left(\frac{eV_{BE}}{k_B T}\right) - 1\right]. \quad (6.35)$$

Here, we have explicitly introduced the negative value of the collector-base bias (using the magnitude signs to account for this). We also

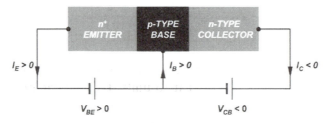

Figure 6.17. Notation used to identify the various currents and applied bias voltages in the *n–p–n* bipolar junction transistor.

have introduced a parameter α, which represents the fraction of the emitter current that reaches the collector (the fraction of the electrons injected into the base that reach the collector-base junction, plus some other corrections to be discussed below). The parameter α is termed the current gain, and is < 1. Since the collector-base bias is usually large compared to the thermal voltage $k_B T/e$, we can ignore the exponential factor in (6.35), and the collector current may be approximated as

$$ I_C \approx -I_{C0} - \alpha I_{E0} \left[\exp\left(\frac{e V_{BE}}{k_B T}\right) - 1 \right]. \qquad (6.36) $$

The quantities I_{C0} and I_{E0} are the reverse saturation currents for these two junctions when they are isolated from the other. That is, these two values depend solely upon the doping in the various regions and the other factors in (6.33).

The term, which is multiplied by α, is simply the total current in the emitter I_E. Thus, we can actually rewrite (6.36) once again as

$$ I_C = -I_{C0} - \alpha I_E. \qquad (6.37) $$

From the circuit of Fig. 6.14, we can also write down the base current as

$$ I_B = (1 - \alpha) I_E - I_{C0}. \qquad (6.38) $$

If we now solve for I_E in this latter equation and substitute it into (6.37), we have

$$ -I_C = \frac{I_{C0}}{1 - \alpha} + \left(\frac{\alpha}{1 - \alpha}\right) I_B. \qquad (6.39) $$

It is the second term which leads to a large current gain in the transistor. If we can make α near to unity, then the factor

$$ \beta = \frac{\alpha}{1 - \alpha} \qquad (6.40) $$

can be made $\gg 1$. For example, if we can make $\alpha = 0.95$, then $\beta = 19$. If we make $\alpha = 0.99$, then $\beta = 99$. Obviously, since we desire to make transistors with large current gains, we want to make α as close to unity as possible.

6.2.2. The Current Gain α

As discussed, the key ingredient of the bipolar transistor is the injection of excess charge into the base from the base–emitter junction. This charge then diffuses across the base, where it varies the reverse current in the collector–base junction. The motion of the charge across the base is governed by the diffusion equation (5.125). In this case, it is written for electrons, as we will continue to treat the n–p–n transistor, so that this equation becomes

$$D_e \frac{\partial^2 \Delta n}{\partial x^2} - \frac{\Delta n}{\tau_n} = 0. \tag{6.41}$$

Contrary to the last chapter, we take the x axis as the direction of current flow, as in the previous sections of this chapter. On the base side of the base-emitter junction, we can write the excess charge as

$$\delta n_p = n_p - n_{p0} = n_{p0}\left[\exp\left(\frac{eV_{BE}}{k_B T}\right) - 1\right]. \tag{6.42}$$

This condition provides one of the two allowed boundary conditions for (6.41). That is, we take the origin at the base side of the base–emitter depletion region, so that

$$\Delta n(0) = \delta n_p. \tag{6.43}$$

For the other boundary condition, we assume that carriers which reach the base–collector junction are swept across the junction, thus contributing to the collector reverse current. Hence, we will assume that

$$\Delta n(W_B) = 0. \tag{6.44}$$

In fact, this is not strictly correct, but it suffices for the accuracy that we desire in this simple treatment of the base transport. The solutions to (6.41) can be written as

$$\Delta n(x) = A e^{-x/L_e} + B e^{x/L_e}. \tag{6.45}$$

This latter is the general solution, and we now can apply our boundary conditions. At $x = 0$, we use (6.43) to give

$$A + B = \delta n_p. \tag{6.46}$$

On the other hand, at $x = W_B$, we use (6.44) to give

$$A e^{-W_B/L_e} + B e^{W_B/L_e} = 0, \tag{6.47}$$

or

$$B = -A e^{-2W_B/L_e}. \tag{6.48}$$

Using this result in (6.45), we find that

$$A = \frac{\delta n_p}{1 - e^{-2W_B/L_e}}. \tag{6.49}$$

We can now substitute these results back into the general solution (6.45) to give

$$\begin{aligned}
\Delta n(x) &= A[e^{-x/L_e} - e^{(x - 2W_B)/L_e}] \\
&= A e^{-W_B/L_e}[e^{-(x - W_B)/L_e} - e^{(x - W_B)/L_e}] \\
&= \delta n_p \frac{\sinh[(x - W_B)/L_e]}{\sinh(W_B/L_e)}.
\end{aligned} \tag{6.50}$$

It can be easily confirmed that this result satisfies both boundary conditions, and thus provides the variation of the excess charge across the base region of the transistor.

The current in the base region can now be found by use of the current equation in the presence of diffusion, given by (6.30). Thus, the electron current in the base is given by

$$J_n = eD_e \frac{d\Delta n(x)}{dx} = -\frac{eD_e \delta n_p}{L_e} \frac{\cosh[(x - W_B)/L_e]}{\sinh(W_B/L_e)}. \qquad (6.51)$$

One component of the current gain α is the fraction of the current that reaches the collector junction, which is termed the base transport factor

$$B = \frac{J_n(W_B)}{J_n(0)} = \frac{1}{\cosh(W_B/L_e)} \sim 1 - \left(\frac{W_B}{L_e}\right)^2. \qquad (6.52)$$

The last expression is valid when the base width W_B is smaller than the electron diffusion length. As mentioned, B is one factor in the current gain α. The desired result is to have electrons injected from the base into the emitter, and these carriers appear as a large collector current. For this to happen, the primary factors are the fraction of the electrons which get across the base, given by B, and the fraction of the base-emitter current that is carried by injected electrons. This second factor is the emitter efficiency

$$\gamma_E = \frac{J_{E,\text{electrons}}}{J_{E,\text{electrons}} + J_{e,\text{holes}}} = \frac{eD_e n_p/L_e}{eD_e n_p/L_e + eD_h p_n/L_h} = \frac{1}{1 + \dfrac{D_h L_e p_{n0}}{D_e L_h n_{p0}}}. \qquad (6.53)$$

The dominant part of this expression is the ratio of the two minority carrier densities. If we want this quantity to be near unity, then we need to have $n_{p0} \gg p_{n0}$, which means that we desire $N_{A,E} \gg N_{D,B}$. That is, we want the emitter heavily doped relative to the base, so that the largest fraction of the emitter current is electrons injected into the base, rather than holes injected into the emitter from the base. Finally, we have the current gain as

$$\alpha = \gamma_E B. \qquad (6.54)$$

If we use the previous values of the parameters for the p–n junction of the base–emitter contact, we have $N_{D,E} = 10^{18} \text{ cm}^{-3}$ and

$N_{A,B} = 10^{17}\,\mathrm{cm}^{-3}$. Then we have $\gamma_E \sim 0.9412$. If the base is 2 μm thick, then $B = 0.9968$, and $\alpha = 0.938$. This gives a $\beta \sim 15$. It is clear that more can be done to improve this parameter, but that the emitter injection efficiency is the primary fact that needs improvement.

One point needs to be emphasized here, and that is the collector doping. We recall from the previous section that the major part of the depletion width is in the lightly doped side of the junction. For the transistor to work properly, the base must remain well defined. This means that we cannot allow the collector depletion width to reach to the base–emitter junction. Hence, since the collector-base junction is reverse biased, we must use a lightly doped collector region. This keeps the depletion width in the collector region, but at the same time introduces series resistance in the device (the current flow in the collector creates a potential drop that differs from the actual bias level). The trade-offs of design of the bipolar transistor include making some compromises in the doping of the collector and the actual width of the base region.

6.3. THE METAL–SEMICONDUCTOR JUNCTION

We have already considered the properties of a junction in which two types of material—n-type and p-type—are brought together to make the junction. What if we replace one of the materials, say the p-type region, with a metal? While the details are different, the resulting behavior is quite similar to the junction diode. However, the result depends, of course, on the doping of the semiconductor material—in essence on which of the two types is replaced. At the end of the last chapter, and even in Chapter 3, we discussed the metal *work function*. In principle, the work function tells us the energy difference between the Fermi energy of the metal and the vacuum level (where the electrons are in vacuum). Similarly, we introduced the work function for the semiconductor with the same description. When the semiconductor is brought together with the metal, and a connection is made to the back side of both, band bending occurs in the semiconductor (and a small amount in the metal). In essence, the two materials must come together in a manner such that the Fermi

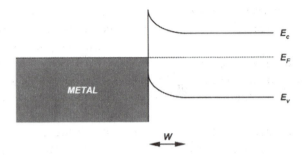

Figure 6.18. When a junction is formed between a metal and a semiconductor, the transfer of charge between the two materials can result in the formation of an energy barrier between them, as we show here. The case shown here is an example of a rectifying Schottky barrier and a crucial feature is the existence of *different* energy barriers for charge transfer from the semiconductor to the metal, and vice versa.

energy is continuous (and not spatially varying) across the junction. This will force some band bending, usually in the semiconductor. Such band bending means that there is a charge depletion in the semiconductor, which is balanced by charge accumulation in the metal (this accumulation accounts for the small band bending in the metal — normally this is unobservable on the scale of band bending in the semiconductor, but it still exists). This is shown in Fig. 6.18. The charge motion actually occurs through the electric circuit of the back contacts, rather than through diffusion across the junction. To understand why this is the case, we must consider how *thermal equilibrium* is established. When the semiconductor–metal structure is isolated, we cannot be sure of anything within the structure. However, when we bring both sides into contact with the "real world" existing at a temperature of 300 K, then certain conditions transpire, and these require the Fermi level at each end to be at a value consistent with the thermal "bath" to which it is connected. If there is no current flow, then both ends are at the same potential, and the Fermi energy must be flat. For this to occur, electrons must leave the semiconductor through the back contact, and they must enter the metal through its back contact. These are not the same electrons, but come from the reservoir that is the thermal bath. It is easier to think of these as the same electrons, since the same amount of charge is involved. The charge cannot just diffuse across the junction, otherwise there would be no barrier. It is the existence of the barrier between the metal and the semiconductor, and no barriers between these two constituents and the bath, that makes everything happen. Similarly, electrons cannot flow from the semiconductor valence band to the metal, as all states in the metal below the Fermi energy are filled. Hence, there is also a barrier for holes as well as for electrons.

The band bending associated with the space charge predominantly occurs in the semiconductor. Normally, the doping in the semiconductor is in the range 10^{17}–10^{18} cm^{-3}. On the other hand, the density of electrons in the metal is of the order of 10^{22} cm^{-3}, which is orders of magnitude larger. If we assume the semiconductor is n-type, and the metal replaces the p-type region, then $x_p \to 0$ due to the large density in the metal. Thus, the entire depletion width is given by x_n in the semiconductor. On the other hand, the peak electric field still occurs at the interface between the metal and the semiconductor.

The charge depletion still creates a potential drop across the junction, which corresponds to the band bending shown in Fig. 6.18. This band bending is completely analogous to the built-in potential of the p–n junction. We plot the variation of the electric field and the built-in potential in the metal–semiconductor system in Fig. 6.19. From (6.18), we can write this built-in potential as

$$V(x) = -\frac{eN_D}{2\varepsilon}(x - x_n)^2 + C_1, \qquad x > 0. \qquad (6.55)$$

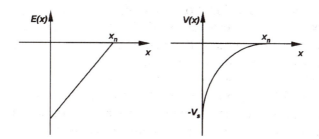

Figure 6.19. The electric field (left) and electrostatic potential (right) variations as a function of position in the metal–semiconductor contact. Note that the depletion region is contained almost exclusively within the semiconductor.

Here, N_D is the doping in the n-type semiconductor, x_n is the width of the depletion region ($=W$), and C_1 is an integration constant as before. Here, however, we have one boundary condition that $V(x_n) = 0$. This leads to $C_1 = 0$. Note that this is a different reference for the potential than that used in the p–n junction. Here, we chose the bulk of the semiconductor as the reference potential, so that the surface potential is written relative to that level. As a result, the surface potential is given as

$$V_s = V(0) = -\frac{eN_D x_n^2}{2\varepsilon}. \tag{6.56}$$

By the same token, the electric field at the surface is given by

$$F_s = \frac{\partial V(x)}{\partial x}\bigg|_{x=0} = \frac{eN_D x_n}{\varepsilon}. \tag{6.57}$$

As in the p–n junction, the application of an applied bias will shift the Fermi energy in the semiconductor relative to that at equilibrium. A forward bias (positive on the metal) reduces this barrier, while a negative bias increases the barrier. The method of the current flow is somewhat different, as current from the metal flows into the semiconductor either by tunneling through the barrier or by thermionic emission over the barrier. Nevertheless, the current flow still is given by

$$I = I_0\left[\exp\left(\frac{eV_a}{k_B T}\right) - 1\right]. \tag{6.58}$$

In a p-type semiconductor, somewhat different behavior is found. In this case, the Fermi energy in the semiconductor is near the valence band edge. As a result, the band bending takes the opposite direction from that shown in Fig. 6.18. The depletion charge also has the opposite sign, as it is holes which are swept out of the depletion region. This changes the sign of the surface potential and the field at the surface, but otherwise the equations have the same general form.

We have used the idea of the work functions to discuss how the built-in potential is formed in the metal–semiconductor junction, which is called a *Schottky-barrier diode*.[4] In fact, in most semiconductors, the barrier height has very little to do with the metal (or the semiconductor) work function. Rather, it is thought that the band bending is dominated by defect levels at the interface between the metal and the semiconductor. This latter model is termed the Bardeen model.[5] Measurements by Kurtin, McGill, and Mead[6] confirmed this behavior for most semiconductors. We can understand the difference between the two models. In the work function model, the band bending is described by

$$eV_s = E_{c,\mathrm{surf}} - E_{c,\mathrm{bulk}} = \Phi_m - \Phi_s - (E_{c,\mathrm{bulk}} - E_F)$$
$$= \Phi_m - \Phi_s - k_B T \ln(N_c/N_D). \tag{6.59}$$

Here, the metal work functions play a major part, with the doping providing a variation around the nominal value. In the Bardeen model, however, there is a range of surface, or interface states, at the junction. These are occupied up to a point at which the Fermi energy is *pinned*. (That is, there is a certain position in the band gap where the density of interface states is so high that it is almost impossible to move the Fermi energy past this point — hence the term *pinning*.) This limits the band bending. In this model, we write the surface potential as

$$eV_{bi} = E_{c,\mathrm{surf}} - E_{c,\mathrm{bulk}} = (E_{c,\mathrm{surf}} - E_{\mathrm{defect}}) - (E_{c,\mathrm{bulk}} - E_F)$$
$$= E_{c,\mathrm{surf}} - E_{\mathrm{defect}} - k_B T \ln(N_c/N_D). \tag{6.60}$$

The change in the first line is to recognize that the pinning occurs with the Fermi energy at the defect level. Here, the critical role of the difference between the two work functions is replaced by the difference between the conduction band edge and the defect level at the interface.

One can ask how the above difference is really determined. In fact, what was measured by Kurtin, McGill, and Mead is the barrier height, for many different metals on a given semiconductor, by a variety of electrical and optical measurements. Then, this barrier height was plotted as a function of the work function of the metal.

They then described the surface potential through the relationship

$$eV_{bi} = S\Phi_m + \text{constant.} \tag{6.61}$$

They found that for most homopolar semiconductors like Si and Ge, $S = 0$. For heteropolar materials such as GaAs, S was slightly larger than 0, but no where near unity. From this result, it was concluded that the Bardeen model is the dominant model. In spite of this, there is no real pinning in Si or Ge, and the Fermi energy can be swept through the band gap under certain conditions (the density of surface/interface states is relatively low). In GaAs (as in most Ga-based compounds), however, the pinning is strong with $E_c - E_{\text{defect}} \sim 0.85$ eV. However, in ionic materials, where there was little covalent bonding, these authors found that S increased toward unity. In fact, they plotted the value of S versus the electronegativity difference of the two materials in the compound $\chi_A - \chi_B$. For a value of this difference less than about 0.4, S was near zero. On the other hand, for a value >0.4, they observed that the value of S rose to unity. However, Michael Schlüter[7] re-examined the data some years later, and suggested that S did not really saturate near unity, but could actually be found to rise to a value of 2 to 3. This behavior is outside either of the two theories, and remains unexplained today, in that this value is *not* given by the work function difference.

6.4. THE SCHOTTKY-GATE TRANSISTOR

From (6.56), it is clear that the built-in potential may be defined as the value of the potential (which describes the band bending) in the bulk minus that at the surface. Hence, we can write this as

$$V_{bi} = V(x > x_n) - V_s = \frac{eN_D x_n^2}{2\varepsilon}. \tag{6.62}$$

Moreover, we can connect the width of the junction as $W = x_n$. How can we use this to make a transistor? In Sec. 6.2, the transistor operated on the control of the collector current by a small base

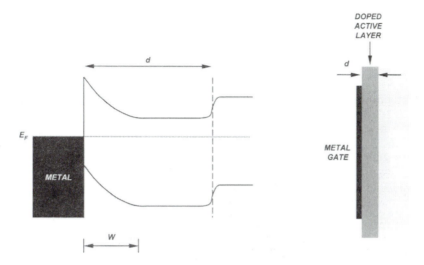

Figure 6.20. Energy-band diagram (left) and basic composition (right) of the MESFET (metal–semiconductor field-effect transistor) device. The picture shown here corresponds to thermal equilibrium, which can be easily confirmed by the invariance of the Fermi level as a function of position.

current. Here, no current really flows through the Schottky barrier when it is reversed bias (in the ideal case). On the other hand, the width of the depletion region can be varied by the applied potential, as

$$W(V_a) = \sqrt{\frac{2\varepsilon[V_{bi} - V_a]}{eN_D}}.$$
(6.63)

We can use this to control the resistance of a semiconductor layer of a finite thickness. Consider Fig. 6.20, where we use a Schottky barrier on top of a thin, epitaxial layer. The bulk material is very low conductivity semi-insulating material (nearly intrinsic at room temperature, so the conductance is quite small). The epitaxial layer is doped sufficiently high, say 10^{17} cm^{-3}, to allow current flow in this layer. The Schottky barrier can now be used to vary the effective thickness of this layer, by depleting the region under the Schottky

barrier — hence the name metal–semiconductor field-effect transistor, or MESFET. The thickness of the epitaxial doped layer is taken
to be d. Then, the resistance of this layer is approximately

$$R_{\text{epi}} = \frac{\rho L}{A} \sim \frac{m^*}{N_D e^2 \tau} \frac{L}{Z(d - W)}, \tag{6.64}$$

where $n \sim N_D$ is assumed, L is the length of the gated region, and Z
is the width of this region. Hence, by varying the applied voltage to
the metal "gate," the thickness of the conducting region can be
varied, which in turn varies the current flow through the region.
Typically, the voltage applied to the gate is negative, so as to avoid
forward current in the junction.

We make a transistor by adding a "source" and "drain" for the
conducting electrons, as shown in Fig. 6.21. Here, the electrons move
from the *source* to the *drain*, hence the terminology, and therefore the
drain-source potential is positive, as indicated. The gate is used to
deplete the channel by narrowing it, so the gate-source voltage is
negative, as indicated. The source and drain regions themselves are
heavily doped regions where the current-carrying contacts can be
placed. The idea is that these regions are doped sufficiently heavily
that the depletion width, with a metal connection, is thin enough that
the electrons can easily tunnel into and out of these regions (from–to
the metal contacts). While the idea of a "contact" is quite simple, its

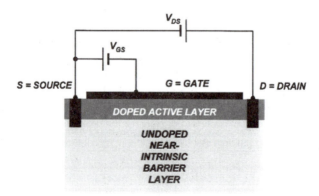

Figure 6.21. Biasing circuit for the MESFET. The source and drain are heavily doped regions
where current-carrying contacts are made.

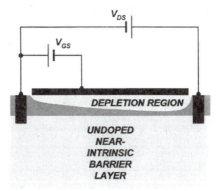

Figure 6.22. By suitable biasing of the gate and source and drain electrodes, the width of the depletion region that forms under the gate in the active layer can be sensitively varied. Note that the thickness of this depletion region varies nonuniformly along the length of the channel, due to the presence of the applied source-drain voltage.

actual implentation is not so simple. The idea works well in Si, but less so in other materials.

Note that the source electrode is taken to be the reference potential for the circuit. Unfortunately, the depletion width under the gate electrode is not uniform along the channel, which is taken as the y axis. It is easy to see why this is the case. At the source end of the gate, the potential across the reverse biased junction is approximately $V_{bi} + V_{GS}$. However, at the drain end of the gate, the total potential across the depletion region is $V_{bi} + V_{GS} + V_{DS}$. That is, the junction is more reverse-biased at the drain end, and hence the channel is narrower at this end. This effect has to be incorporated into any description of the current flow through the structure. We carry this out with reference to Fig. 6.22. We take the potential to be zero at the source end of the channel (and gate). This is also taken as the point $y = 0$. At the drain end of the channel, $y = L_g$ and the potential is V_{DS}. At any point $0 < y < L_g$, the potential drop is given by a linear resistance variation as

$$V(y) = \frac{R_{0y}}{R_{0y} + R_{yL_g}} V_{DS} = \frac{R_{0y}}{R_{0L_g}} V_{DS}. \tag{6.65}$$

The resistance varies with the thickness of the channel $(d - W)$, so the voltage drop at point y is given by

$$dV(y) = I \cdot dR(y) = I \frac{\rho \, dy}{Z(d-W)} = I \frac{dy}{ZN_D e\mu_e [d - W(y)]}. \quad (6.66)$$

Note that Kirchhoff's current law requires that the current be continuous throughout the structure. We can now rearrange this equation to give the form

$$I \, dy = ZN_D e\mu_e [d - W(y)] \, dV(y)$$
$$= ZN_D e\mu_e \left[d - \sqrt{\frac{2\varepsilon[V_{bi} + V_{GS} + V(y)]}{eN_D}} \right] dV(y). \quad (6.66)$$

In the last line, we have used the width from (6.63), noting that the *applied* voltage is composed of both the negative gate bias and the positive channel potential, and is the value $-[V_{GS} + V(y)]$. Equation (6.66) is a simple differential equation, which is valid so long as $d - W > 0$. If we are to integrate this equation, we have to decide upon boundary conditions, particularly for the voltage on the right-hand side of the equation. We have already discussed these as

$$\begin{aligned} V(y) &= 0, & y &= 0, \\ V(y) &= V_{DS}, & y &= L_g. \end{aligned} \quad (6.67)$$

Using these boundary conditions, we can integrate (6.66) to yield (6.68)

$$I = \frac{ZN_D e\mu_e}{L_g} \left\{ dV_{DS} - \frac{2}{3}\sqrt{\frac{2\varepsilon}{eN_D}} [(V_{bi} + V_{GS} + V_{DS})^{3/2} - (V_{bi} + V_{GS})^{3/2}] \right\}.$$

$$(6.68)$$

For small values of the drain–source bias potential, we can expand the power law terms to obtain the approximate *linear* result

$$I = \frac{ZN_D e\mu_e}{L_g} [d - W(0)]V_{DS}. \quad (6.69)$$

Hence, the current is varied directly by the width of the junction at the source end. In fact, the conductance of the channel is what is being varied. The channel is "pinched off" when the depletion width reaches all the way across the channel, or

$$W(0) = d = \sqrt{\frac{2\varepsilon(V_{bi} + V_{GS})}{eN_d}}. \qquad (6.70)$$

Suppose we use Si for the epitaxial layer. Here, the mobility is about 1200 cm^2/Vs for a doping concentration of 10^{17} cm^{-3}. For a Au gate, the surface potential is about 0.6 V, so that the built-in potential can be found from (6.59) as 0.455 V. This gives a depletion width of 77.7 nm. If we now use an epitaxial layer thickness of 0.25 μm, then this thickness is "pinched off" for an applied $V_{GS} = -4.25$ V. When the gate source voltage is zero, the channel conductance for a 1.0 μm gate length is found from (6.69) to be 0.33 S/mm (that is, 0.33 S for each mm of width Z of the channel in the z direction). This is usually stated as 330 mS/mm.

As the drain voltage is raised, the first term in the square brackets of (6.68) can cause the current to decrease. In fact, the derivative of the current with respect to the voltage goes to zero when

$$W(L_g) = d = \sqrt{\frac{2\varepsilon(V_{bi} + V_{GS} + V_{DS})}{eN_d}}. \qquad (6.71)$$

Equation (6.68) is no longer valid for drain voltages above that required to achieve this condition:

$$V_{DS,\text{sat}} = \frac{eN_D}{2\varepsilon}[d^2 - W^2(0)], \qquad (6.72)$$

which is termed the *saturation* voltage. Equation (6.68) can now be manipulated, using (6.70) and (6.72) to give the saturation current

$$I_{\text{sat}} = \frac{ZN_D e\mu_e}{L_g} \frac{eN_D}{2\varepsilon} \left[\frac{d^3}{3} - dW^2(0) + \frac{2W^3(0)}{3} \right]. \qquad (6.73)$$

In the saturation regime, the current is no longer a function of the drain potential, but is varied only by the gate-source voltage. A set of output curves is shown in Fig. 6.23. At saturation, the depletion region reaches entirely through the epitaxial layer at the drain end of the channel. This does not, however, cut off the current due to the overwhelming importance of the Kirchhoff current law. Current is flowing through the source end of the channel, and has to go

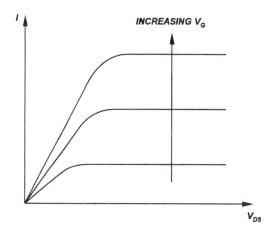

Figure 6.23. Typical current–voltage characteristics for the MESFET and their dependence on gate voltage. The most-negative gate voltage is the lowest curve and subsequent curves are for reducing the magnitude of this voltage.

somewhere. Where it goes is the breakdown of the theory. We have assumed only drift current in setting of the potential drop. In fact, the carriers are injected into the depletion region, just as in a $p–n$ junction diode, and then diffuse to the drain. The current is determined by the properties at the source end of the channel, so it is this input end that is the most important part of the device. In Fig. 6.24, we show a scanning electron micrograph of a GaAs MESFET with a channel length of 35 nm. The epitaxial layer is only 60 nm thick, and doped to 3×10^{18} cm^{-3}. The device has been fabricated with electron-beam lithography, and gives gain up to 165 GHz.

6.5. THE METAL-OXIDE–SEMICONDUCTOR FIELD-EFFECT TRANSISTOR (MOSFET)

The diodes and transistors, which we have studied above, have predominantly depended upon a $p–n$ junction (or metal–semiconductor junction) for their operation. In the transistors, it was the bias across one of the $p–n$ junctions which determined the current flow in the device. In the bipolar transistor, it was bias across the base–

GATE

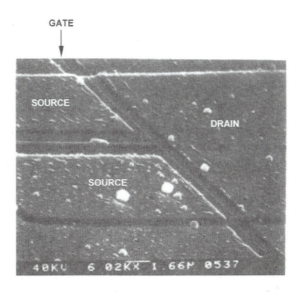

Figure 6.24. An electron micrograph of a GaAs MESFET. The source is split, with the gate feed passing between the two sides. This is a common feature of microwave devices.

emitter junction which set the collector current. In the MESFET, it was the bias across the metal–semiconductor junction which set the channel opening and the consequent source–drain current.

There are other semiconductor devices in which an insulator is placed on the surface, so that this insulator forms the central part of a metal-oxide–semiconductor (MOS) structure. In these materials, such as Si, the surface position of the Fermi energy is not strongly pinned and can be moved relatively easily. For example, in a MOS structure, such as sketched in Fig. 6.25, a positive voltage on the gate will pull electrons to the surface of the semiconductor. On the other hand, a negative voltage on the gate will push electrons away, and pull holes to the surface. If we use a p-type semiconductor, then a positive gate bias can even *invert* the surface to n-type material, by pulling a sufficient number of electrons to the surface. Hence, we can create a p–n junction at the surface of the semiconductor. The desirability of this MOS structure arises from two factors: (1) the ease of moving carriers at the surface to create different properties, and (2) the advantageous properties of the native oxide of Si—silicon dioxide, SiO_2. Silicon dioxide is one of the most common materials

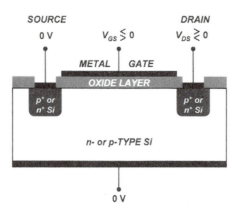

Figure 6.25. Schematic illustration of a MOSFET. The notation for the relevant voltages is also defined in this figure. The "upper" notation is used for p-channel devices, while the "lower" notation is used for n-channel devices.

in nature, but it is one of the best insulators known in electronics. The fact that very high quality SiO_2 can be grown by simply putting a Si wafer in a high temperature furnace is exceedingly important to the manufacturing of modern VLSI. It was this easy growth of SiO_2 that was used in Sec. 6.2 for fabricating the $p-n$ junction by various stages of photo-lithography. We shall see below that this MOS structure can be used to create a field-effect transistor in a manner much like the MESFET described above. However, the fact that we can make both n-channel and p-channel devices (in Si) makes the MOS field-effect transistor (MOSFET) a far more desirable device structure for integrated circuits. Here, the channel is actually induced by the MOS structure, and this, in turn, provides the modulation of the conductance between a source and a drain that creates the transistor action.

6.5.1. The MOS Structure and the Surface Channel

Let us consider a p-type semiconductor material, as shown in Fig. 6.26(a). When the semiconductor material is connected to the metal by a "back" contact (as in the metal–semiconductor structure), the Fermi level must line up so that it is invariant with position. Thus,

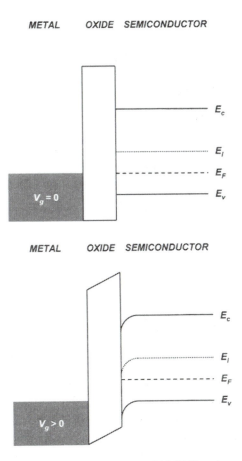

Figure 6.26. (a) The energy bands in an ideal p-type MOSFET under conditions of thermal equilibrium (no applied gate voltage). The Fermi level is constant throughout the device. (b) Application of a positive bias to the gate bends the energy bands in the oxide and semiconductor regions. When this bending is sufficiently strong the intrinsic Fermi level may drop below the Fermi level in the semiconductor, causing *inversion* of the majority carriers close to the interface.

the Fermi level in the metal gate must align itself with that in the semiconductor. In the MESFET, the interface charge often was sufficiently great that it dominated the zero-bias condition and "pinned" the Fermi level in the semiconductor at the surface. In the MOS structure, we require semiconductors which have very good surfaces, as defined by very low numbers of surface (or interface when the oxide is grown) states. In GaAs, where the MESFET is most commonly found, the density of interface states is more than 10^{12} cm^{-2}, which approaches the atomic density at the surface. On the other hand, it is no problem, with proper processing, to reduce this density by more than two orders of magnitude at the $Si-SiO_2$ interface. This means that, for the latter MOS structure, one can actually move the Fermi energy at the interface relatively easily.

When the gate is biased positively with respect to the semiconductor, electrons are drawn toward the interface with the oxide. At first, these electrons recombine with holes, leaving only the negatively charged ionized acceptors in a narrow region adjacent to the interface, as shown in Fig. 6.26(b). For still larger values of the positive gate bias voltage, still more electrons are drawn to the interface, which at first broadens the depletion width throughout which the holes have been removed. In both cases, the conduction and valence bands are bent at the interface to account for this change in the number of free holes. Finally, at a critical bias voltage, known as the "turn-on" voltage, electrons begin to accumulate at the interface, creating an n-type region with mobile electrons. That is, we have *inverted* the type of the majority carrier at the surface, from holes to electrons.

Just where do these electrons come from? If we have no other contacts than that of the gate and the back of the p-type material, then the electrons must be thermally generated. This is a very slow process, and the resultant modulation of the surface channel is also very slow. However, if we create n-type regions at either end of the MOS structure (also on the surface of the semiconductor; these are the *source* and *drain* regions), then electrons can easily flow out of these regions to populate the surface channel that is induced by the MOS voltage. That is, in the absence of the MOS voltage applied to the gate, the two n-type contacts and the p-type semiconductor create an $n-p-n$ structure with a very long base "width," which is a poor transistor. However, if a positive voltage is applied to the MOS gate, then an n-type channel can be formed between the two n-type contact regions. The number of electrons, and therefore the conductance of the channel, is easily modulated by the MOS voltage, and this creates a field-effect transistor (FET)—a MOSFET. Such MOSFETs are the heart of modern integrated circuits, and more than 40 million of these devices may be found on the latest Pentium chips.

The "source" and "drain" contacts—named precisely as they were in the MESFET—look like reverse-biased $p–n$ junctions with respect to the p-type semiconductor which forms the base structure of the transistor. The action of the gate produces an n-type interface layer, which *inverts* the surface (changes it from p-type to n-type). This allows current flow between the source and drain, and the size of this current is easily modulated by the gate voltage. At first, the gate bias depletes the holes from the surface, creating a depletion region at the interface. For higher gate voltage, finally electrons create the inversion region for which the n-type channel is turned "on." The voltage at which this latter occurs is called the "turn-on" voltage. In Fig. 6.27, we plot the charge in the interfacial region as a function of the gate voltage. For forward bias, we move first through the depletion region, and then into the inversion region. On the other hand, for negative gate voltage, we only pull more holes to the surface, creating a larger hole population at the interface, termed *accumulation*. Current cannot flow easily in this configuration due to

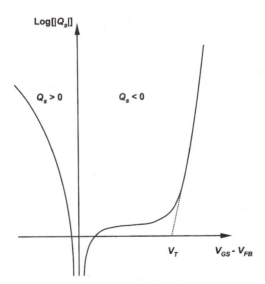

Figure 6.27. Variation of the surface charge density as a function of gate voltage in the MOSFET. Note that the charge density plotted for negative gate voltage is an accumulation charge, while for positive gate voltages larger than the threshold value (V_T) inversion of the majority carriers occurs at the interface. The curve is shown for a p-type substrate.

the reverse bias on the source and drain $p-n$ junctions. As in the previous section, we will discover that a set of transistor curves can be generated, with linear behavior at small V_{DS}, and saturation occurring at larger V_{DS}.

6.5.2. The MOSFET Characteristics

To begin to understand how the MOSFET works, we first have to correct the picture of Fig. 6.26(a). It is highly unlikely that the Fermi energy in the semiconductor will *exactly* match the position of the Fermi energy in the metal throughout the structure. A variation can occur due to charges in the oxide, as well as just due to variations of the Fermi level in the semiconductor with doping. As a consequence, one must actually apply a small voltage to the gate to bring the bands in the semiconductor into the flat position shown in Fig. 6.26(a). The amount of voltage that must be applied is known as the *flat-band voltage* V_{FB}. When this voltage is applied, then there is no net space charge at any point *in* the semiconductor p-type region. Any changes to this voltage causes a charge to accumulate at the interface between the semiconductor and the oxide. If the voltage is positive on the gate, then the charge is negative (electrons are pulled to the surface). If the voltage is negative, then the charge is positive (electrons are pushed away from the surface and holes are pulled to the surface). In general, we take the voltage at the source as the reference potential, and refer to V_{GS} as the voltage on the gate relative to the source. Hence, we can write the charge that accumulates at the semiconductor–oxide interface as

$$-Q_{\text{int}} = C_{\text{ox}}[V_{GS} - V_{FB}]. \tag{6.74}$$

It must be pointed out that the charge Q_{int} is actually the surface charge density in C/m^2. The oxide capacitance per unit area C_{ox} is $\varepsilon_{\text{ox}}/t_{\text{ox}}$, where t_{ox} is the thickness of the oxide layer and ε_{ox} is the permittivity of the oxide (approximately $3.4\varepsilon_0$ in SiO_2). We use the *specific* quantities (values per unit area) as we do not yet know the area of the oxide capacitor.

Once $V_{GS} > V_{FB}$, negative charge begins to accumulate at the interface between the oxide and the semiconductor. This charge is composed of two components. One is the ionized acceptors which are no longer compensated by the holes, as the latter have been pushed away. The second component is the electrons which have been drawn to the interface by the band curvature, shown in Fig. 6.26(b). In both cases, the actual charge at any position along the channel will depend upon the *local* voltage, just as in the MESFET. However, this MOSFET construction is an *inversion* device. We use the gate to change the surface from *p*-type to *n*-type — we *invert* the surface type. In the MESFET, we used the gate to push charge out of the channel, or to deplete the channel. Hence, the MESFET is a *depletion* device. It is the good properties of the oxide, and its interface with Si, that allows inversion devices to be made with desirable characteristics.

Consider Fig. 6.28, in which we show the structure of an actual MOSFET with source and drain, and a drain bias. The problem we have is that the charge at the source end of the channel is given by (6.74), while the charge at the drain end of the channel is given by

$$-Q_{\text{int}} = C_{\text{ox}}[V_{GS} - V_{FB} - V_{DS}]. \tag{6.75}$$

That is, the charge is determined by the potential between the gate and the drain. With a large gate voltage, the band bending at the source end of the channel should look like Fig. 6.26(b). On the other hand, if $V_{DS} = V_{GS} - V_{FB}$, the band bending at the drain end of the channel will look like Fig. 6.26(a). In this latter case, there is no

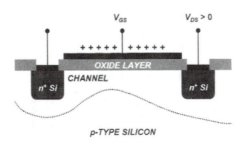

Figure 6.28. A schematic diagram illustrating the nonuniform thickness of the electron channel as a function of distance from the source to the drain contacts. The nonuniform thickness is a consequence of the similarly nonuniform depletion regions that develop around the source and drain regions and the variation in the surface potential in the channel.

channel at the drain, and the current is saturated just as in the MESFET. To calculate a simple expression for this current, we have to know just how much band bending exists at each point along the channel. However, we are really interested in the amount of the mobile charge, due to the electrons.

The problem is approached precisely as in the case of the MESFET in the last section. We will develop an equation for the small local voltage drop at an arbitrary point along the channel in terms of the current and the conductance at that point. This will then be integrated over the channel length to get a single expression for the current in terms of the terminal voltages indicated in Fig. 6.25. There is one difference here, and that is we deal with the sheet resistance R_s, sheet conductance G_s, and sheet density n_s (electrons per square meter) of electrons in keeping with the surface charge density Q_{int} defined in (6.74). Hence, we can write the incremental sheet resistance dR_s at an arbitrary point y as

$$dR_s = \frac{1}{Z\sigma_s} dy = \frac{1}{Zn_s e\mu_e} dy. \tag{6.76}$$

Here, Z is the width of the MOSFET in the direction normal to the plane of Fig. 6.28. Once the gate voltage is sufficiently large, electrons begin to accumulate at the interface, and their sheet density is given by

$$en_s = C_{ox}[V_{GS} - V_{FB} - V_T] \tag{6.77}$$

at the source end of the channel. The difference between this result and (6.74) is that we have introduced the turn-on voltage and deal only with the electronic contribution to the surface charge. Here, V_T is the turn-on voltage introduced in the last section. Away from the source end, the surface potential is somewhere between the zero value at the source and V_{DS} at the drain. Thus, at our arbitrary point y, we must modify (6.77) to

$$en_s(y) = C_{ox}[V_{GS} - V_{FB} - V_T - V(y)]. \tag{6.78}$$

We can write the voltage drop at the point y as

$$dV(y) = I\,dR_s = \frac{I\,dy}{Zn_s(y)e\mu_e}. \tag{6.79}$$

Rearranging this equation, and inserting (6.78), we arrive at

$$I\,dy = ZC_{ox}\mu_e[V_{GS} - V_{FB} - V_T - V(y)]\,dV(y).$$ (6.80)

Before integrating (6.80), it is useful to introduce a reduced potential which combines the flat-band and turn-on voltages, as

$$V_T' = V_{FB} + V_T,$$ (6.81)

and this is usually referred to as the *threshold* voltage. Now, the limits on the integration of (6.80) are obviously $y = 0$ and $V(y) = 0$ at the source end of the channel, and $y = L_g$ and $V(y) = V_{DS}$ at the drain end of the channel. The integration can now be easily performed to give

$$\begin{aligned}
I &= \frac{ZC_{ox}\mu_e}{2L_g}\left\{\left[V_{GS} - V_T'\right]^2 - \left[V_{GS} - V_T' - V_{DS}\right]^2\right\} \\
&= \frac{ZC_{ox}\mu_e}{L_g}\left[V_{GS} - V_T' - \frac{V_{DS}}{2}\right]V_{DS}.
\end{aligned}$$ (6.82)

This expression for the drain current is valid so long as the term in the square brackets is positive. As in the earlier case of the MESFET, saturation in the current occurs for a sufficiently large V_{DS}. We can differentiate (6.82) with respect to V_{DS}, and then set this derivative to zero to find the saturation value of the drain voltage as

$$V_{DS,sat} \equiv V_{sat} = V_{GS} - V_T'.$$ (6.83)

For voltages above this value, the current is saturated at the value

$$I_{D,sat} = \frac{ZC_{ox}\mu_e}{2L_g}(V_{GS} - V_T')^2.$$ (6.84)

In Fig. 6.29, we plot a set of characteristics with increasing gate voltage. The assumption that the mobility μ_e is a constant is the source of the error. In fact, the mobility *decreases* with increasing gate voltage due to increasing interface scattering, which lowers the mobility. Thus, when we plot the current as a function of the drain

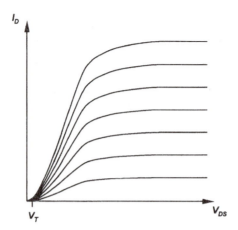

Figure 6.29. Typical current–voltage characteristics for an n-channel MOSFET and their dependence on gate voltage (which is positive in this case).

bias, with gate bias as a parameter, we do not see the quadratic behavior of (6.84) either.

The condition at which saturation sets in, as given by (6.83), is exactly the value discussed above for which the inversion density disappears at the drain end of the channel. Once the electrons get to this point they are injected into the depletion region, and subsequently move to the drain by a combination of drift and diffusion. The current remains governed by (6.84), as this is still the determining condition, which is set near the source end of the channel. That is, the current level is largely set by the properties of the local potential at the source end of the channel, but saturation is initiated by drain end properties. This is a general property of all field-effect transistors, here as well as in the MESFET discussed earlier. We can estimate the size of the currents from the above equation, particularly (6.82). In the year 2000, the primary gate length being used for front line microprocessors was 0.25 μm. If the oxide thickness is 25 nm, then $C_{ox} = 1.2$ mF/m^2. If we operate with 1.5 V for V_{GS} and say 0.2 V for V_{DS} (the normal operation in logic circuits will have a low source-drain voltage for a high gate voltage, as we will see in a subsequent section). Usually then, the threshold voltage will be approximately 0.3 V, so that we can estimate the sheet density n_s as 9×10^{11} cm^{-12}. The field in the oxide is approximately 6×10^5 V/cm, for which the mobility will be about 500 cm^2/Vs. We can now compute the current density as about 53 mA/mm. If the drain voltage were higher, so that the device was in saturation, then the current would be 173 mA/mm, or about a factor of 3 higher. It is clear that the conditions for logic operation make the device operate more likely in the "linear" regime than in the saturated regime. On the other hand, this is the steady-state, and it is clear that the drain voltage is high when the gate potential is first applied.

6.6. THE HIGH-ELECTRON-MOBILITY TRANSISTOR

It has always been recognized that GaAs would be a more desirable material than Si, because the mobility is as much as a factor of 10 higher. As can be seen from (6.84), this would give a factor of 10 higher current, if the capacitance could be maintained. However, the problem is that the oxides of GaAs are terrible for device application. The very properties that yield the Fermi level pinning in the metal–semiconductor junction are those which preclude having a good MOS behavior. We cannot get around the fact that nature has blessed Si devices with an almost perfect oxide and interface. Nevertheless, we can create an "insulator" by clever materials processing. In Fig. 4.19, we sketched the interface between AlGaAs and GaAs. The conduction band offset between these two materials leads to significant band bending at the interface. Consider a layer of AlGaAs grown on top of lightly p-doped GaAs, and then a metal gate placed on the AlGaAs layer, as shown in Fig. 6.30. As usual, the Fermi energy must be continuous throughout the layer, as no current flows through the structure. This leads to the metal–semiconductor barrier behavior at the interface between the metal and the AlGaAs. The heterostructure band bending is also shown at the interface between the AlGaAs and the GaAs. For the moment, we have indicated that the AlGaAs is uniformly doped with donors to create an n-type material. The band bending at the M-S interface depletes the electrons, and the space charge due to the ionized donors is balanced by negative charge on the metal. On the other hand, the band bending at the heterointerface causes the electrons to be depleted

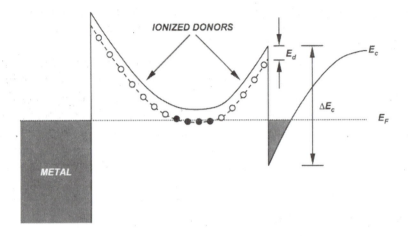

Figure 6.30. Energy band diagram at thermal equilibrium in a HEMT (high electron mobility) structure. A high-mobility electron channel forms at the interface between the two semiconductors, where the discontinuity in the band gaps results in the formation of a potential well.

from the AlGaAs in this region as well, and these electrons accumulate in the quantum well on the GaAs side of the interface, as shown. In the central region of the AlGaAs, the bands are flat and there is a free electron concentration in the conduction band that is given by the donor density (since the donors are ionized — the Fermi energy is below the donor level).

As the structure stands, it makes a terrible device. One wants to use the gate to control the carrier density in the quantum well that exists on the GaAs side of the heterointerface. The fact that the Schottky-barrier depletion does not reach through the AlGaAs prevents this from happening. In addition, free carriers in the AlGaAs are undesirable, as they have low mobility and degrade the device performance. Hence, one wants the thickness of the AlGaAs layer d to be less than the combined thickness of the two depletion regions in this material. We can then overcome both problems — we deplete the AlGaAs and use negative gate voltage to deplete the channel in the GaAs quantum well. This makes a depletion-mode device. On the other hand, if the thickness d is sufficiently small, the Schottky barrier will automatically deplete the channel, and forward bias can be used to induce charge into the channel. This is called an *enhancement-*

mode device (it is not called an inversion device, since the carriers come from the donors in the AlGaAs, although this difference is rather a hair-splitting difference).

In Fig. 6.31, a negative gate voltage has been applied to the metal, with respect to the GaAs. In the latter material, there is no additional band bending, so there is no charge at the interface. All the potential is across the depletion region of the Schottky barrier. The Fermi energy in the metal, and that in the GaAs are separated by the applied negative gate voltage $\Delta E = -eV_G$. The energy difference between the Fermi energy in the metal and the conduction band edge of the AlGaAs at this interface is denoted as ϕ_m, which is an effective Schottky barrier height — it differs from the built-in potential by the donor ionization energy, as

$$\phi_m = -eV_{bi} - E_{di}, \qquad E_{di} = E_{c,\text{AlGaAs}} - E_{d,\text{AlGaAs}}. \qquad (6.85)$$

We can write an Kirchhoff potential loop equation to give

$$-eV_G + E_{di} = \Delta E_c + (E_{c,\text{GaAs}} - E_{F,\text{GaAs}}). \qquad (6.86)$$

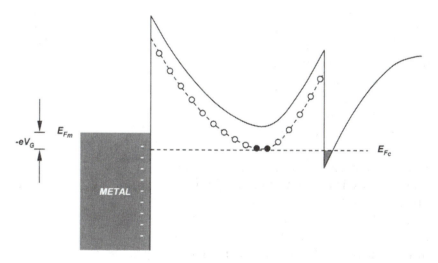

Figure 6.31. Energy band diagram for the HEMT structure of Fig. 6.30, this time for the situation where a negative voltage is applied to the gate. Note how the negative voltages causes further ionization of the donors and also raises the potential of electrons at the interface of the two semiconductors. Ultimately, a sufficiently large negative voltage may be used to deplete the electron channel completely. We have introduced the notation E_{F_c} and E_{F_m} here to denote the position of the Fermi level in the semiconductor and metal, respectively.

The term in the parentheses can be replaced by the doping properties of the lightly p-type GaAs, and the equation rearranged to give

$$-eV_{G,T} = -E_{di} + \Delta E_c + E_{G,\text{GaAs}} - k_B T \ln\left(\frac{N_v}{N_A}\right) \equiv -eV_T. \quad (6.87)$$

This value of gate voltage is the *threshold* voltage for the channel, and is determined primarily by the properties of the GaAs and the donor ionization energy in the AlGaAs. In most situations, the GaAs is semi-insulating, or very lightly p-doped, so that the last two terms on the right combine for a value of about 0.7–0.8 eV. However, the situation shown is one in which the thickness is just the proper amount that the built-in potential completely depletes the carriers in the AlGaAs.

In any case, at this point any more positive gate voltage begins to accumulate charge in the channel just as in the MOSFET. Hence, this charge creates an additional field in the AlGaAs, which is terminated in the additional charge on the metal, both of which are determined by the change in the gate voltage. As a result, we can talk about the "inversion" charge density as being given by

$$en_s = C_{\text{AlGaAs}}[V_{GS} - V_T]. \quad (6.88)$$

We recall that both voltages are negative, with $V_{GS} > V_T$, so the right-hand side of this equation is positive. At this point, the development proceeds exactly as in the MOSFET case, with the same results as that of (6.82) and (6.84). The only differences are that the values of the AlGaAs dielectric constant and the mobility of GaAs are used in the two equations for the current.

As mentioned, the thickness of the AlGaAs layer has to be balanced with the doping to produce the correct band bending. For example, if the Fermi energy pinning position in the AlGaAs is 0.8 eV ($= \phi_m$), then the built-in potential has to be less than this amount. The built-in potential for the Schottky barrier is given by (6.56) as

$$eV_{bi} = \frac{e^2 N_D d^2}{2\varepsilon}. \quad (6.89)$$

Hence, we require the thickness of the AlGaAs and the doping to fit the criteria that

$$\frac{eN_D d^2}{2\varepsilon} < 0.8 - \frac{E_{di}}{e}. \qquad (6.90)$$

We remark, of course, that one can use this last equation to replace the poorly known value of E_{di} in (6.87) and therefore achieve a better estimate of the threshold voltage. It is generally thought that this donor ionization energy is somewhere between 60 and 100 meV. Let us take the worst case of 0.1 eV, and assume that the material is doped to 10^{18} cm^{-3}. Then, (6.90) tells us that d is of the order of (or less than) 30 nm. For $x \sim 0.33$, the conduction band discontinuity is about 0.3 eV, and the threshold voltage is approximately -1 V. The maximum available sheet charge in the device is given by the product $N_D d \sim 3 \times 10^{12}$ cm^{-2}, but the most charge that can be put into the channel is less than a third of this. To produce the same charge as the MOSFET example above, we need to have $V_{GS} - V_T \sim 0.35$ V, due to the larger dielectric constant of the AlGaAs. The mobility in GaAs is about 7500 cm^2/Vs in the channel, which is higher than that of Si, as used in the preceding. For a drain voltage of 0.1 V, we then find from (6.82) that we obtain a current of 370 mA/mm, for the same 0.25 μm gate length. This is almost a factor of 7 increase in the current available from the device, so this would be a good n-channel FET. Why aren't they used more? The problems is that most logic applications are achieved today with complementary devices (coupled n- and p-channel devices), and this is a very difficult technology to create in GaAs-based structures (it has never really gotten out of the laboratory). As a result, these devices have been limited to very high frequency (microwave) applications rather than logic applications, although this is changing. The gate patterns shown in Fig. 6.24 have also been used to make HEMTs, so this image could as easily be of a HEMT.

Because of the ease of depleting the channel at the interface between the GaAs and the AlGaAs, these surface gates can also be used to make novel quantum structures. For example if we simply make two gates with a small opening between them, as shown in Fig. 6.32. At low temperatures, we may use the surface potential to create

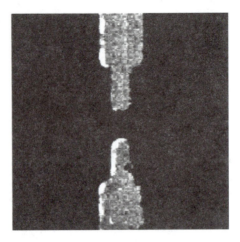

Figure 6.32. A picture of the two metal gates that create a quantum point contact. The narrow gap between the two gates is just a few hundred nanometers wide.

lateral barriers whose width is comparable to the electron wavelength. The carrier density in the interface is mainly two dimensional, and we can adapt (5.27) to have

$$n_s = \frac{2}{(2\pi)^2} \int\!\!\int f_{FD}\, dk_y\, dk_z = \frac{2}{(2\pi)^2} \int_0^{2\pi} d\varphi \int_0^{k_F} k\, dk = \frac{k_F^2}{2\pi} = \frac{2m^*E_F}{2\pi\hbar^2}.$$

(6.91)

In this equation, we have assumed that all states up to the Fermi energy are filled (low temperature) and all states above the Fermi energy are empty. Moreover, we require the Fermi energy to lie in the conduction band—the material is heavily degenerate. Using (6.79), we can rewrite the current as

$$I = Zn_s e\mu_s \frac{dV_y}{dy} \sim Zn_s e\mu_e \frac{V_y}{L},$$

(6.92)

where L is the length of the quantum point contact. However, we must make another change, and that is to the mobility. In (6.92), the mobility is used with a relaxation time τ_e from (5.67). Here, we want the transport to be *ballistic*, in that no scattering occurs. Hence, we will replace τ_e with the transit time L/v_y, where v_y is the velocity of carriers at the Fermi surface. Now, (6.92) becomes

$$I = \frac{Zn_se^2}{m^*v_F}V_y, \tag{6.93}$$

and the conductance becomes

$$G = \frac{I}{V_y} = \frac{Zn_se^2}{m^*v_y} = \frac{Zn_se^2}{\hbar k_F}. \tag{6.94}$$

In a quantum structure, we make the width $Z = N\lambda_F/2$, where N is an integer. That is, we make the width an integer number of half-wavelengths at the Fermi energy. In order to account for the Z, however, we modify the integration in (6.91) to account for the wavelength limitations. Hence, (6.91) becomes

$$n_s = \frac{1}{2\pi^2}\int_{-N\pi/2Z}^{N\pi/2Z}dk_z\int_{-k_F}^{k_F}dk_y = \frac{Nk_F}{\pi Z}. \tag{6.95}$$

When, we now substitute the above expressions for Z and n_s into (6.94), we find

$$G = N \cdot \frac{2e^2}{h}. \tag{6.96}$$

That is, the conductance through our little quantum point contact of Fig. 6.32 is quantized into units of $2e^2/h$, a factor of 2 difference from the quantization found in the quantum Hall resistance in the last chapter (where h/e^2 was the quantum Hall resistance). In Fig. 6.33, we show measurements of this effect in a GaAs–AlGaAs heterostructure at low temperature.

The use of surface gates can be used to fabricate a variety of quantum structures, all of which are measured at low temperatures so that the quantization becomes significant in comparison to the thermal energy k_BT. Another typical structure is the quantum dot, shown in Fig. 6.34. Here, a sizable region constrained by lateral potentials (induced by the gates) is connected to the rest of the inversion layer by two quantum point contacts.

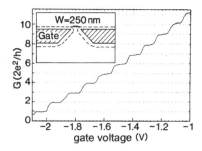

Figure 6.33. Measurements of the quantized conductance through a similar quantum point contact to that shown in Fig. 6.32. *B. J. van Wees et al. Phys. Rev. Lett.* **60**, *848 (1988). Figure reproduced with permission.*

6.7. COMPLEMENTARY MOS STRUCTURES

Before ending this chapter, it is worthwhile to undertake a brief study of circuits to see how the MOSFET fits into the logic gate, and how this leads to the fabrication of the gate. We chose the complentary MOS (or CMOS) gate, as it is the backbone of modern VLSI. In

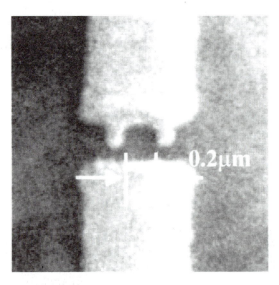

Figure 6.34. A quantum dot formed by connecting a small region through two quantum point contacts. The white spacer bar in this picture denotes a distance of 1 μm.

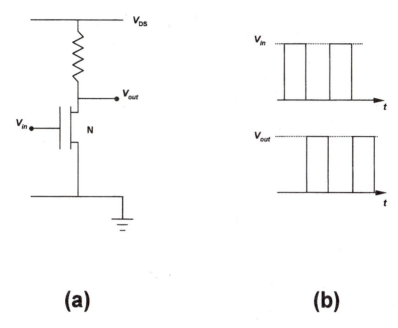

(a)   **(b)**

Figure 6.35. (a) Circuit diagram for the *n*-MOS resistor inverter. (b) Input–output characteristics for the inverter.

principle, it is composed of a pair of MOSFETs, one *n*-channel and one *p*-channel, which are connected in a manner that one is always on and the other is always off. To understand how this occurs, let us consider a simple *n*-channel transistor switch, in which the transistor is connected to the drain supply V_{DD} through a resistor, as shown in Fig. 6.35(a). The input signal is connected to the gate of the transistor, while the output is taken at the drain terminal. When a positive voltage is applied to the gate, the transistor is switched into the "on" state, for which the drain voltage drops rapidly, due to the *IR* drop in the resistor. This is shown schematically in Fig. 6.35(b). The transistor goes into saturation and the output voltage drops very near to zero. Thus, the output is "inverted" relative to the input, but the power dissipated in the resistor is quite high, and is a significant disadvantage for this type of circuit.

Now, consider the *p*-channel circuit shown in Fig. 6.36(a). In essence, the source is at the power supply terminal V_{DD}, while the drain is connected to the output terminal. Again, the input terminal

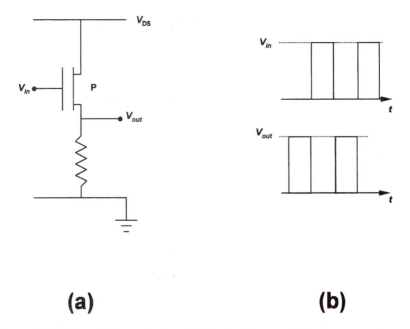

(a)   **(b)**

Figure 6.36. (a) Circuit diagram for the p-MOS resistor inverter. (b) Input–output characteristics for the inverter.

is connected to the gate of the transistor. When the input voltage goes to a "low" state (low voltage), there is a large gate-source potential drop, which pulls holes to the surface and turns "on" the p-channel device. This causes current to flow, and the output goes to a "high" state due to the IR drop in the resistor. While the operation is different from that of the n-channel device, the result is the same. A "low" input voltage gives a "high" output voltage, and a "high" input voltage gives a "low" output voltage, as shown in Fig. 6.36(b). Again, the power dissipated in the resistor is quite high, and is a significant disadvantage for a circuit like this.

6.7.1. The Complementary Circuit

The genius of the complementary circuit is to replace each of the resistors with a transistor of the opposite type. Thus, in Fig. 6.35(a), the resistor is replaced with a p-channel device. In Fig. 6.36(a), the

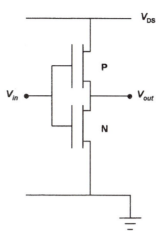

Figure 6.37. Circuit diagram for the CMOS inverter. This yields the same inverter action as the devices of Figs. 6.30 and 6.31 but dissipates significantly less power.

resistor is replaced with an n-channel device. In either case, we obtain a resulting circuit in which there are two transistors in series. The p-channel device is connected between the supply voltage V_{DD} and the output terminal, while the n-channel device is connected between the output terminal and ground. The two gates are tied together, as shown in Fig. 6.37. Again, when the input goes "high," the n-channel transistor is turned "on," and the p-channel transistor is turned "off." The output terminal is pulled to the ground level, but very little current flows through the off p-channel device, so there is very little dissipation. On the other hand, when the input goes "low," the n-channel transistor is turned "off," and the p-channel transistor is turned "on." The output terminal is pulled up toward V_{DD}, but there is little current flow as the n-channel device is off. The current through the p-channel device can flow only into the load, but if this is another gate, this current is quite small. Thus, the CMOS invertor circuit dissipates very little power in any situation except during the switching process, when both transistors can be on.

In Fig. 6.38, a possible material structure for the CMOS gate is depicted. Each group who builds such structures has its own design and layout. The one shown in Fig. 6.38 is a conceptual one that doesn't necessarily bear any resemblance to one actually used in the

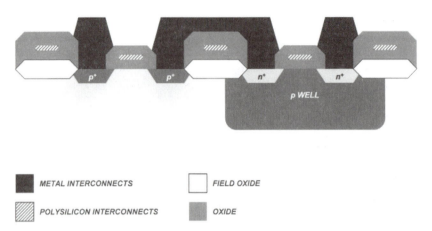

Figure 6.38. A schematic diagram of a CMOS inverter.

industry. Here, life begins with a lightly doped n-type substrate, although either type could be used. Then, two processing sequences are performed to create n-type wells and p-type wells (or *tubs*, as they are often called), in which the transistors will be placed. A process sequence would entail (1) growth of an oxide to protect the substrate, (2) a lithographic step which opened a window where the well is to be placed, and (3) a doping process, either ion implantation or diffusion, to set the doping level in the well. The remnants of these processes are designed to leave the *field oxides* (which provide isolation), shown in the figure. Then another processing sequence would be used to define the gate oxides and to deposit the poly-Si gates, and to then pattern these gates to provide the interconnects of the circuit. The poly-Si lines on top of the field oxides are not part of the gate circuit, but are interconnects passing through the cell from one side to the other. The source and drain dopings of the two transistors are then accomplished by two separate processing sequences. Finally, the metal interconnects are set down by a final processing sequence. In an actual integrated circuit, it will not be uncommon to have six or seven levels of interconnect, although far fewer are shown in Fig. 6.31. In small devices, it is not uncommon to encounter leakage between the n-channel and p-channel devices. One cure for this is to place a deep (vertical in the figure) cut between the two transistors and to fill this "trench" with oxide for isolation.

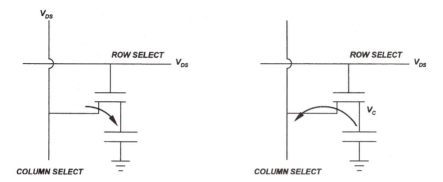

Figure 6.39. A DRAM circuit. (Left) Writing of the memory cell. (Right) Reading of the memory cell.

6.7.2. The DRAM Cell

A variation of the above process is used to create the dynamic, random-access memory cell that is ubiquitous in the computer industry. The circuit is shown in Fig. 6.39. Here, when V_{DD} is applied to both one terminal of the MOSFET and to the gate, current flows from the power supply lines into the capacitor, charging it to V_{DD}. When the voltage is applied only to the gate, then the capacitor charge will flow out through the transistor to the column select line. Thus, we read into the capacitor, Fig. 6.39(left), when both lines are "high," and we read out of the capacitor when only the row select (in this circuit) is high, Fig. 6.39(right).

The layout of such a memory circuit is a modification of the CMOS circuit of Fig. 6.38, in which the *p*-channel device is replaced by the capacitor. This is shown in Fig. 6.40. While the *p*-channel device could be used as the capacitor, space is saved by making the capacitor vertical. Here, the trench isolation is modified by placing a metal line in the center of the trench, which creates a trench capacitor. The problem that arises as the structure grows smaller, through continued integration, is that the trench occupies too much real estate on the chip. Many manufacturers have taken to putting the capacitor on top of the transistor, and using a high dielectric constant material to allow the capacitor to be made smaller. We remark that a modern microprocessor contains a great deal of memory as well as logic circuits.

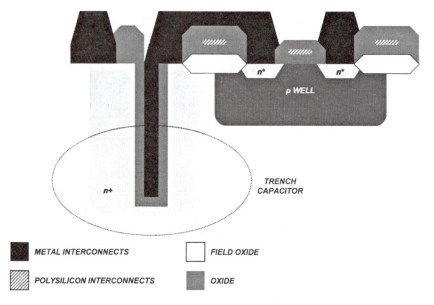

n⁺

TRENCH CAPACITOR

METAL INTERCONNECTS

FIELD OXIDE

POLYSILICON INTERCONNECTS

OXIDE

Figure 6.40. A conceptual DRAM integration using a trench capacitor for the storage.

REFERENCES

1. C. Zener, Proc. Roy. Soc. (London) **145**, 523 (1934).
2. J. Bardeen and W. Brattain, Phys. Rev. **74**, 230 (1948).
3. W. Shockley, Bell Sys. Tech. J. **28**, 435 (1949).
4. W. Schottky, Naturwissenschat. **26**, 843 (1938).
5. J. Bardeen, Phys. Rev. **71**, 717 (1947).
6. S. Kurtin, T. C. McGill, and C. A. Mead, Phys. Rev. Lett. **22**, 1433 (1969).
7. M. Schlüter, J. Vac. Sci. Technol. **15**, 1374 (1978).

PROBLEMS

1. Show that the built-in potential in a $p-n$ junction is given by

$$V_{bi} = \frac{k_B T}{e} \ln\left(\frac{n_n}{n_p}\right) = \frac{k_B T}{e} \ln\left(\frac{p_p}{p_n}\right).$$

2. Using the expression in the previous problem, and the properties of Si: $\mu_e = 1400$ cm^2/Vs, $\mu_h = 400$ cm^2/Vs, $\sigma_n = 1$ mho/cm, $\sigma_p = 100$ mho/cm, and $n_i = 10^{10}$ cm^{-3}, find the built-in potential in a junction with these properties.

3. Assume a p–n junction made from silicon with $N_D = n_n = 10^{16}$ cm^{-3}, $N_A = p_p = 10^{17}$ cm^{-3}. Using the parameters for Si that have been given previously, calculate n_i, n_p, p_n, V_{bi}, W, and the peak electic field. Assume a step junction of area 0.1 cm^2. What is the capacitance of the junction?

4. For a Si junction doped with $N_D = 10^{16}$ cm^{-3}, $N_A = 10^{18}$ cm^{-3}, and area of 0.015 cm^2, calculate the current carried by electrons and holes for a forward bias of 0.15 V. What is the total current?

5. Consider a Si diode of area 1 mm^2 and doping of $N_A = 10^{15}$ cm^{-3} and $N_D = 10^{17}$ cm^{-3}. Plot the current as a function of the applied bias.

6. A Si junction doesn't go into breakdown until the electric field reaches 6×10^5 V/cm. If $N_D = N_A = 10^{17}$ cm^{-3}, calculate the width and applied voltage at the onset of avalanche breakdown.

7. A particular semiconductor is characterized with $n_i = 7 \times 10^{10}$ cm^{-3} and a dielectric constant of 11.8. the material is doped with $N_D = 10^{16}$ cm^{-3}. At one end, the material is counterdoped with $N_A = 10^{18}$ cm^{-3}. What are the built-in potential and the width of the junction?

8. Consider a Si bipolar transistor with $N_{A,E} = 10^{18}$ cm^{-3}, $N_{D,B} = 10^{17}$ cm^{-3}, and $N_{A,C} = 10^{16}$ cm^{-3}. If the junction area is 1 mm^2, calculate the characteristics and the value of β for the device. Assume a base width of 5 μm, and $\tau_p = \tau_n = 1$ μs.

9. For an n-channel MESFET whose channel width is 10 μm, and which has a doping of 10^{16} cm^{-3} and a doped layer thickness of 5 μm, determine the pinch-off voltage. Assume the Schottky barrier height is 0.8 V and the material is GaAs. If the channel is 0.1 μm long, calculate the transconductance

$$g_m = \frac{\partial I_D}{\partial V_{GS}}$$

in saturation.

10. An n-channel GaAs Schottky-gate FET has an active layer thickness of 100 nm, a channel width of 5 μm, and a gate length

of 0.2 μm. If the Schottky barrier potential is 0.75 V, what is the pinch-off voltage for a doping of 10^{18} cm^{-3}?

11. Sketch the energy band diagrams and the charge distribution in an MOS structure, fabricated on n-type material, under biasing conditions corresponding to carrier accumulation, depletion, and inversion. Neglect the role of surface states and work function differences.

12. Derive expressions of the bulk charge (the net charge of the non-compensated acceptors when the holes are pushed away), surface potential, and surface electric field for an MOS structure at the onset of strong inversion (when the Fermi level is as close to the conduction band at the surface as it is to the valence band in the bulk). Plot these quantities as a function of the acceptor doping concentration in the bulk in the range of 10^{14} cm^{-3} to 10^{18} cm^{-3}.

13. An n-channel MOSFET has an oxide thickness of 100 nm, oxide dielectric constant of 4, $Z/L_g = 10$, $\mu_e = 1000$ cm^2/Vs, and $V_T = 0.5$ V. Calculate the saturation current. Assume $V_{GS} = 5$ V.

14. An n-channel GaAs–AlGaAs HEMT has an AlGaAs thickness of 300 nm (dielectric constant of 16). If the latter material is uniformly doped to 10^{18} cm^{-3} and a donor ionization energy of 0.1 eV, the electrons transferred to GaAs may be determined from the interfacial electric field. This field is the equivalent of that produced in one side of a p–n junction if we determine an equal number of ionized donors in the AlGaAs. At the edge of the depletion region in the AlGaAs, we move from the compensated (by free electrons) to the uncompensated donors. Thus, the width of the depletion region is given by

$$x_d^2 = V_{bi} \frac{2\varepsilon}{eN_{Di}}.$$

At the same time, the field is given by

$$E_{int} = \frac{eN_{Di}x_d}{\varepsilon}.$$

At the point $x = x_d$, the Fermi level resides at the donor level. If the inversion charge in the GaAs is 10^{12} cm^{-2}, what is the built-in potential and the width of the depletion region on the AlGaAs side of the heterojunction?

CHAPTER 7

Dielectric Material

In previous chapters, we discussed an assortment of materials with varying dielectric constants. What gives these materials these particular properties? Why does one material have a dielectric constant of 4 and another 16? While we cannot answer this question in detail, we can address what leads to the dielectric constant, and how this property can be used for a variety of electronics applications. In this chapter we will endeavor to do just this.

7.1. DIELECTRIC EFFECTS

A simple parallel-plate capacitor consists of no more than two plates of area A separated by a dielectric material of thickness d. In treating such a capacitor, we find that the surface charge depends on the applied voltage and that the quantity that measures this is the capacitance. In the simple parallel-plate capacitor, the capacitance is given by

$$C = \frac{\varepsilon A}{d}, \qquad (7.1)$$

where ε is the *permittivity* and is given by

$$\varepsilon = \varepsilon_r \varepsilon_0, \qquad (7.2)$$

where ε_0 is the permittivity of free space and is 8.854×10^{-12} farads/m, and the voltage and charge are related by the form

$$Q = CV. \tag{7.3}$$

It has been demonstrated experimentally that if some material, such as an *insulator*, is placed between the plates of the capacitor, the capacitance is raised. In the preceding treatment, to accommodate this observable fact, we have introduced a different permittivity in (7.2), through the relative dielectric constant, or the relative permittivity of the insulator. If the capacitance is raised by inserting the insulator between the plates, then the voltage across the capacitor must be reduced for a constant charge on the plates. But the voltage is just

$$V = -\int_0^d F \, dx, \tag{7.4}$$

where the x direction is taken normal to the plates. Therefore, we must conclude that the electric field between the plates is reduced by the insulator, even though the charge on the plates remains the same.

Now how can this occur? For, if we think about it, it is a peculiar enough phenomenon. To be sure, since the electric field depends strongly on the charge density on the plate surface, a reduction of this field must imply that the charge seen from within the dielectric must be less than the actual surface charge. One way that this could happen is for the dielectric to have charge induced on its own surface in such a way that the negatively charged capacitor plate draws positive charge to the surface of the dielectric (we saw this effect in the MOS capacitor of Chapter 6). This occurs in a conductor due to electrical conduction. Although it is not exactly what occurs in an insulating dielectric, it does yield an obvious model of what happens. The dielectric can be considered to be composed of a number of small separated spherical conductors. Then each one has a surface charge density that contributes to the total charge of the material.

It turns out, however, that the small regions of conductors are not necessary at all. The surface charge tends to make the small spherical conductors act like a dipole, that is, like two separate and opposite charges at opposite sides of the sphere. The only important fact is that these dipoles exist, not where they come from. The dipoles in fact are caused by the atoms and electrons themselves.[1] This dipole polarization can occur by polarization of the electron cloud about the nucleus, or even from a polarization of the lattice itself. A little later we shall discuss some other sources of polarization, but at present let us just consider the case in which the electron cloud about each nucleus is polarized by the electric field.

If an electric field is applied to an atom, then the positively charged nucleus tends to move in the direction of the electric field, while the electrons move in the opposite direction. The extent to which these move is limited by the inter-atomic forces, but a slight displacement is induced (Fig. 7.1). If the electric field is not too large, then the displacement should be proportional to the field. A small field gives a small displacement; a large field gives a larger displacement. In each atom, a small displacement is created. Suppose that the charges q are separated a distance x_0; then we can define a *dipole moment* per atom as

$$d_i = qx_0. \tag{7.5}$$

If there are n atoms per unit volume, the net dipole moment per unit volume is nqx_0. The dipole moment may be represented by the polarization

$$P_x = nqx_0 \tag{7.6}$$

along the direction of F and x_0. The polarization P_x is proportional to the electric field F. The constant of proportionality depends on the ease with which the electrons are shifted, hence on the nature of the

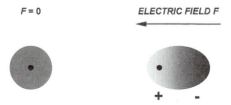

Figure 7.1. When an electric field is applied to an initially unpolarized atom (shown left) an internal polarization of the atom develops (shown right). Note that this polarization gives rise to an internal electric field that opposes the externally applied field.

atoms and solid. The polarization represents a set of charges separated by a small distance. Charges separated in this manner create electric fields. A consideration of Fig. 7.1 shows that this new field is in a direction that opposes, or reduces, the applied field F, just as is needed to increase the capacitance and reduce the net internal electric field.

In the solid dielectric, consider what happens at the surfaces. At one surface, near the positive plate of the capacitor, the electrons have effectively moved out a distance x_0. At the other surface, they have moved inward, leaving a positive charge. Thus there will be a net surface charge density on the dielectric. This surface charge is called the *polarization charge*. Let us say that the surface area is A; then the surface charge is

$$Q_{\mathrm{pol}} = (\text{no. of charges per unit vol.}) \times \text{vol.} \times q. \qquad (7.7)$$

We find the surface charge per unit area by dividing by A and

$$\rho_{s,\mathrm{pol}} = nqx_0 = P_x \ \mathrm{C/m^2}, \qquad (7.8)$$

so that the surface charge density is equal to the polarization, a not unexpected result.

If there exists a surface charge, it may be recalled that

$$C = \frac{Q}{V} = \frac{\rho_s A}{F_s d} = \frac{\varepsilon_0 A}{d}, \qquad (7.9)$$

where

$$F_s = \frac{\rho_s}{\varepsilon_0}, \qquad (7.10)$$

ρ_s is the surface charge density, and F_s is the electric field due to this charge. Thus, the surface charge of the dielectric produces an electric field

$$F_{\mathrm{pol}} = \frac{\rho_{s,\mathrm{pol}}}{\varepsilon_0}, \qquad (7.11)$$

and the total electric field in the material is

$$F = F_s - F_{\mathrm{pol}}, \qquad (7.12)$$

where F_s is the field due to the capacitor plate charge if no dielectric is present, and the minus sign has been chosen since the polarization field opposes the applied field. Then, combining (7.10) through (7.12), we find that the net field is

$$F = \frac{\rho_s - \rho_{s,pol}}{\varepsilon_0} = \frac{\rho_s - P_x}{\varepsilon_0}. \tag{7.13}$$

Equation (7.13) doesn't yield a value for F unless we know a value for P_x, which is proportional to F (or F_s), and we may write the proportionality as[2]

$$P_x = \chi_e \varepsilon_0 F, \tag{7.14}$$

where χ_e is called the *electric susceptibility*. Then the net electric field is

$$F = \left(\frac{\rho_s}{\varepsilon_0}\right) \frac{1}{1 + \chi_e}. \tag{7.15}$$

We recognize that the *relative* permittivity is just

$$\varepsilon_r = 1 + \chi_e, \tag{7.16}$$

and that the capacitance is

$$C = \frac{A}{d} \frac{\rho_s}{F} = \frac{(1 + \chi_e)\varepsilon_0 A}{d} = \frac{\varepsilon_r \varepsilon_0 A}{d}, \tag{7.17}$$

which is just the result in (7.1).

The behavior that leads to the susceptibility χ_e can be found for ions, or, in solids, for the atoms held together in a potential interaction, giving an a contribution for the *ionic* polarizability. Of course, this case arises only in polar material, such as GaAs or GaN—materials in which the valence of the two atoms differs. The difference in charge on the two atoms gives an ionic, or Coulomb, contribution to the bonding, as discussed in Chapter 2, particularly for the zinc-blende materials. In addition, the valence electrons also

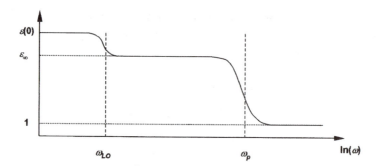

Figure 7.2. Frequency dependence of the dielectric constant, showing the important characteristic frequency ranges.

contribute to the polarizability with their own characteristic frequency. We illustrate this in Fig. 7.2. Figure 7.2 shows the total dielectric constant for a typical material. In this material, there are contributions from electronic and ionic or molecular polarizability. The valence electrons can respond very easily to high-frequency electric fields, so their cutoff is in the ultraviolet region. For this reason, the dielectric constant which includes the electronic contribution is termed the optical dielectric constant ε_∞, and this is the most commonly encountered value for, for example, silicon and other nonpolar materials. In the zinc-blende, and other polar materials, the lattice polarization also contributes below the characteristic optical phonon frequency, which is a few tens of meV. Thus, the low frequency dielectric constant $\varepsilon(0)$ includes the electronic and the lattice contributions to the polarization.

7.1.1. Lattice Polarization

In Chapter 4, we discussed the vibrations of the two different atoms in a diatomic chain. There the equations of motion were written in (4.71) as (we take only the $q \to 0$ result)[3]

$$-\omega^2 M_1 u_s = 2C(w_{s+1} - u_s),$$
$$-\omega^2 M_2 w_{s-1} = 2C(u_s - w_{s-1}).$$
(7.18)

In the case of polar material, an additional force, due to the Coulomb interaction between the two atoms acts on the material, so that these equations become

$$-\omega^2 M_1 u_s = 2C(w_{s+1} - u_s) + e^*F,$$
$$-\omega^2 M_2 w_{s-1} = 2C(u_s - w_{s-1}) - e^*F, \tag{7.19}$$

where e^* is the effective charge on each atom. Since the material is primarily still covalent, only a small amount of charge is transferred from one atom to the other corresponding to the average four electrons on each atom, even though one belongs to group III and one belongs to group IV. We note that the field is opposite on the two atoms. These two equations can be solved to give

$$u_s = \frac{(e^*/M_1)}{\omega_{TO}^2 - \omega^2} F, \qquad w_{s-1} = \frac{(e^*/M_2)}{\omega_{TO}^2 - \omega^2} F, \tag{7.20}$$

where

$$\omega_{TO}^2 = 2C\left(\frac{1}{M_1} + \frac{1}{M_2}\right) \tag{7.21}$$

is the square of the optical frequency we found in (4.75). It is this *transverse* optical mode frequency that corresponds to the diatomic model used in Chapter 4. Here, the presence of the effective charge will lead to a second frequency, the *longitudinal* optical mode frequency, and it is the splitting between these two which gives the ionic contribution to the dielectric constant.

Although the model used is quite simple, it has rather general applicability. The displacement (from the equilibrium position) of the two atoms gives rise to a polarizability as

$$P = \frac{N}{2} e^*(u_s - w_{s-1}) = \frac{S\varepsilon_\infty}{\omega_{TO}^2 - \omega^2} F, \tag{7.22}$$

where

$$S = \frac{Ne^{*2}}{2\varepsilon_\infty}\left(\frac{1}{M_1} + \frac{1}{M_2}\right) = \frac{Ne^{*2}\omega_{TO}^2}{2C}. \tag{7.23}$$

TABLE 7.1. Relative Dielectric Constants of Some Semiconductors

	ε_∞	$\varepsilon(0)$		ε_0	$\varepsilon(0)$
Si	11.9	11.9	GaAs	11.1	13.1
Ge	15.9	15.9	GaSb	14.4	15.7
AlP	11.56	14.88	InP	9.61	15.7
AlAs	8.16	10.06	InAs	11.8	14.5
AlSb	9.9	11.2	InSb	15.7	17.9
GaP	8.5	10.2	HgTe	6.9	14.9
CdTe	7.2	10.2			

Hence, we may write the dielectric permitivitty as

$$\varepsilon(\omega) = \varepsilon_\infty \left(1 + \frac{S}{\omega_{TO}^2 - \omega^2} \right). \tag{7.24}$$

An important, and surprizing result, is that $\varepsilon(\omega) < 0$, and no waves can propagate, for frequencies in the range

$$\omega_{TO}^2 < \omega^2 < \omega_{TO}^2 + S \equiv \omega_{LO}^2. \tag{7.25}$$

Hence, we have defined the longitudinal optical mode frequency ω_{LO} in terms of the transverse optical mode frequency ω_{TO} and the polarization constant S. The splitting between the longitudinal and optical mode frequencies is a direct result of the effective charge and the polarization that results. In nonpolar material, such as Si and Ge, $S = 0$, and the two modes have the same frequency. We give some typical dielectric constants in Table 7.1.

7.1.2. Electronic Polarizability

One of the more common types of polarizability is electronic polarizability, which was discussed earlier when we were determining the susceptibility and dielectric constant. When an atom is placed in an electric field, its electron cloud is shifted somewhat. The amount of shift depends on the nature of the atomic forces and the electric field. The displacement of this electron cloud or distribution is called *electronic polarization*, and we can readily calculate the amount of this shift. Of course, the electrons and atoms of interest here are those within the semiconductor material, so that these are the atoms in the lattice and the electrons are the bonding electrons, which we call the valence electrons — arising from the four electrons per atom that form the covalent bonds.

Recall that the electrons reside in a potential well, given by the atoms themselves. For small movements of the electrons about their equilibrium position, the restoring force is quadratic in nature, much like a particle on a spring. Then it may be said that

$$m\frac{d^2x}{dt^2} = \text{Force} = -m\omega_0^2 x, \tag{7.26}$$

where $m\omega_0^2 = K$ is the "spring constant" of the motion and ω_0 is a characteristic frequency $\sqrt{K/m}$ of the motion. Thus either ω_0 or K is a measure of the atomic potential seen by electrons, and is easily determined, for example, in an isolated Bohr atom (it is somewhat more difficult in a crystalline solid, but we return to this later). If an additional electric force is applied, (7.26) becomes

$$m\frac{d^2x}{dt^2} + m\omega_0^2 x = eF. \tag{7.27}$$

If it is assumed that F varies as $e^{-i\omega t}$, and that consequently, x has this variation as well, then

$$x = \frac{eF}{m(\omega_0^2 - \omega^2)}. \tag{7.28}$$

From (7.28), it is apparent that for $\omega \ll \omega_0$, the displacement is constant and relatively small, while for $\omega \gg \omega_0$, the displacement further decreases rapidly with increasing frequency. For small frequencies, the field is *screened* by the internal interaction. At $\omega = \omega_0$, there is a resonance, and this approach is not valid. Here, the low-frequency case $\omega \ll \omega_0$ is primarily of interest. Then, the dipole moment at each electron around the atom is given by

$$p = ex = \frac{e^2 F}{m\omega_0^2}. \tag{7.29}$$

Multiplying this by the number of valence electrons, we find that the electronic susceptibility due to this interaction may be found from (7.14) to be

$$\chi_{e,\text{electronic}} = \frac{Ne^2}{m\varepsilon_0\omega_0^2} = \frac{\omega_P^2}{\omega_0^2}. \tag{7.30}$$

Here,

$$\omega_P^2 = \frac{Ne^2}{m\varepsilon_0} \tag{7.31}$$

is the *plasma frequency* of the valence electrons. For Si, the inter-atomic spacing is 0.2346 nm, so that there are approximately 5.0×10^{22} atoms/cm^3 and 2×10^{23} valence electrons/cm^3. This gives a valence plasma frequency ω_P of 2.53×10^{16}, or 16.67 eV (the measured value is 16.6 eV), which is well into the ultraviolet region. In GaAs, the interatomic spacing is 0.245 nm, so that there are approximately 4.4×10^{22} atoms/cm^3 and 1.77×10^{23} valence electrons/cm^3. This gives a valence plasma frequency ω_P of 2.37×10^{16}, or 15.6 eV.

To determine the value of ω_0 is somewhat more difficult. This is a characteristic frequency associated with the excitation of the electrons. At frequencies higher than this value, the absorption of the optical photons breaks up the collective motion of the electrons and then destroys this screening effect. This is obviously caused by excitation of the electrons out of the valence states into conduction states. But the crucial energy is not simply the observed optical gap, since the electron can come from anywhere in the valence band and go to anywhere in the conduction band. In fact, the proper value for ω_0 is associated with an "average band gap" between the bonding valence states and the antibonding conduction states. This is closely related to the difference between, for example, the values E_1 and E_2 introduced in Sec. 4.4. While one might actually be able to calculate the value for this important parameter from the band structure, it is more normally determined from the dielectric constant itself via

$$\varepsilon_\infty = \varepsilon_0 \left[1 + \left(\frac{\hbar\omega_P}{E_{G,\text{av}}} \right)^2 \right], \qquad (7.32)$$

where $E_{G,\text{av}} = \hbar\omega_0$. The form (7.32) was determined by Penn,[4] and is often referred to as the Penn dielectric function.

7.2. PIEZOELECTRIC EFFECTS

In materials in which there is no reflection symmetry, it is possible to have a *piezoelectric* interaction. By no reflection symmetry, we mean materials such as the zinc-blende crystals of Fig. 2.16 where each crystal site has a basis of two, and this basis has two different atoms. Examples are GaAs and AlAs, where the basis is composed of the molecular pair Ga–As or Al–As. When we sit at a point midway between these two atoms, there is no inversion symmetry, as we see Ga atoms nearby in one direction, but As atoms nearby in the opposite direction. When we have this dissimilar behavior, an applied electric field interacts with the polar dipole between the two atoms (discussed in Sec. 7.1) and can change the distance between these two atoms. This creates a strain in the crystal which changes the distance between the two atoms. That is, an electric field produces a *distortion* of the lattice, while an induced lattice strain can produce an internal electric field (which is the converse effect).[5]

Another common piezoelectric material is the polymer poly-vinylidene flouride, which is used in a variety of consumer headsets. Application of an acoustic electric field produces a distortion of the film that generates sound waves — the classic action of a loudspeaker. One of the most useful piezoelectric materials is quartz. Pressure applied to the quartz creates an electric field that can be amplified and fed back to the quartz pressure creation mechanism. The advantage of quartz is that there is a crystal resonance, which is controlled by the crystal plane of the material and the thickness of the film. The result is a very stable oscillator, in which the frequency is governed by the internal resonance in the quartz crystal.

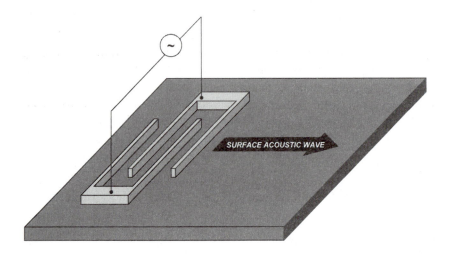

Figure 7.3. A schematic illustration of a surface-acoustic-wave device. A high-frequency voltage is applied to the transducer gates, generating surface acoustic waves.

One modern use of the piezoelectric effect is in surface acoustic wave devices,[6] shown in Fig. 7.3. An *interdigital transducer* creates an electric field on the surface of the device in a manner in which the spatial direction of the field oscillates along the transducer. This couples to an acoustic wave at the surface of the material, since the electric field creates lattice distortion through the piezoelectric effect. The frequency of propagation of the wave along the surface is governed by the acoustic lattice vibrations — the acoustic surface velocity. This velocity is much slower than the speed of light, so these devices have significant application in microwave filters — the inter-digital transducer creates the frequency according to its physical shape.

Finally, one important application of piezoelectric materials is as electrostrictive elements. Electrostrictive elements are those in which the applied electric field produces a noticable elongation (or shorten-ing) of the material — in essence this is the piezoelectric distortion taken to macroscopic levels. One example of the use of these materials is found in the scanning-tunneling microscope (STM) of Fig. 3.16. Since the current that flows between the STM tip and the surface is exponential in the separation of these, measurement of the current is not a good method of estimating the height of the STM from the surface. Instead, this current is measured and used to control a voltage on an electrostrictive element that maintains the tip at a constant distance from the surface. That is, the height is continuously adjusted to maintain a constant tunneling current. Since the displacement of the electrostrictive element is usually linear in the applied voltage, this latter can be monitored as a linear indication of the height of the surface. A change of the surface height, as the tip is scanned, causes the voltage applied to the electrostrictive element to change accordingly, so that the tip is maintained a constant distance from the surface.

7.3. FERROELECTRIC MATERIAL

Certain crystals are permanently polarized, so that $P \neq 0$ even when the applied electric field is zero. These materials are called *ferroelectrics*. In Fig. 7.4, we plot the polarization as a function of the applied electric field. In a ferroelectric, a residual polarization exists even when the electric field has been removed.[7] Moreover, if an oppositely directed electric field, of sufficient magnitude, is applied, this polarization can be reversed. The new direction is maintained after the

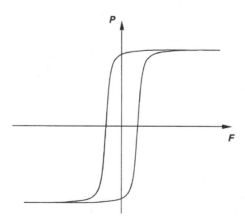

Figure 7.4. The variation of polarization as a function of electric field for a typical ferroelectric material.

electric field is removed. The polarization usually results from a distortion of the crystal structure. Moreover, there is a critical temperature, above which the effect disappears. Consider Fig. 7.5. In the left panel, we show the crystal structure of $BaTiO_3$ for a temperature above the *Curie temperature*, which is the critical temperature. The four corners of the unit cell are occupied by Ba^{2+} ions, whereas the faces of the cube are occupied by O^{2-} ions. In the center of the cube sits a single Ti^{4+} ion. For these high temperatures, the structure is basically cubic. However, the crystal is paranoid as it is not clear whether it wants to be body-centered cubic or face-centered cubic. At high temperatures, the lattice is soft enough that the lattice essentially moves between these two descriptions.

When the temperature is lowered below the Curie temperature, the uncertainty in desired crystal structure results in a distortion of the lattice, shown in Fig. 7.5, right panel. The crystal is elongated along the vertical (termed the *c* axis) direction, and the resulting distortion of the crystal is termed *tetragonal* distortion. If this were all that happened, then we would have nothing new in the crystal. However, the Ti atoms do not accept such a distortion, and they move in a manner to maintain their distance from the top surface of the cell near to that of the high temperature phase. This moves the Ti atoms away from the center of the cell, and creates a built-in

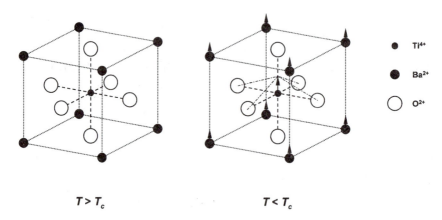

$$T > T_c \qquad\qquad T < T_c$$

Figure 7.5. The crystal structure of barium titanate below (left) and above (right) the Curie temperature (T_c). Below the Curie temperature, a distortion develops with the oxygen ions experiencing a relative displacement with the barium and titanium ions.

polarization along the c axis. On the other hand, the oxygen atoms move smoothly and average the new strain. This displaces the Ti atoms above the oxygen plane. The displacement of the Ti atoms from the center of the cell is denoted as δ_i.

In general, we can determine the ability of the crystal to create the built-in polarization by finding the local (internal) electric field. The polarization within the crystal can be written as

$$\mathbf{P} = \sum_i \mathbf{p}_i = \sum_i e\boldsymbol{\delta}_j = \sum_i \alpha_i \mathbf{F}_{\text{local}}. \tag{7.33}$$

In the last expression, we have introduced the local polarizability, which relates the dipole strength to the local (applied) electric field that results from the distortion of the lattice. The local electric field can also be written as

$$\mathbf{F}_{\text{local}} = \mathbf{F}_{\text{appl}} + \frac{\mathbf{P}}{3\varepsilon_0} = \mathbf{F}_{\text{appl}} + \frac{\varepsilon_r - 1}{3}\mathbf{F}_{\text{appl}} = \frac{\varepsilon_r + 2}{3}\mathbf{F}_{\text{appl}}, \tag{7.34}$$

where the factor of 3 arises from averaging over the various directions. The polarization can be in any direction, as there is no preferred direction for the c axis with respect to our observer coordinates. Now, we can rewrite the susceptibility as

$$\chi = \frac{P}{\varepsilon_0 F_{\text{appl}}} = \varepsilon_r - 1 = \frac{\varepsilon_r + 2}{3\varepsilon_0}\sum_i \alpha_i, \tag{7.35}$$

where we have introduced the polarization from (7.33). Rearranging the last two terms gives the Clausius–Mosotti relation

$$\frac{\varepsilon_r - 1}{\varepsilon_r + 2} = \frac{1}{3\varepsilon_0}\sum_i \alpha_i, \tag{7.36}$$

which, in turn, leads to

$$\varepsilon_r - 1 = \frac{\displaystyle\sum_i \alpha_i}{3\varepsilon_0 - \displaystyle\sum_i \alpha_i}. \tag{7.37}$$

In general, most materials are such that the polarizability α_i is small. However, in ferroelectrics, it is not only not small, but is sufficiently large that the denominator of (7.37) can be driven through zero! Hence, a permanent dipole can exist in the material. In addition, the dielectric constants of the material are usually quite large, even being > 1000. Moreover, the materials are usually anisotropic in that the dielectric constant depends on the crystalline direction. This makes them good for nonlinear optics as well.

7.4. PYROELECTRIC EFFECTS

As a general rule, all ferroelectrics are also piezoelectric, but the converse is obviously not true. In addition, since the distortion and dipole strength vary with the temperature, these materials are also usually pyroelectric,[8] a property to which we now turn. In ferroelectric materials, the permanent dipole (and the large dielectric constant) disappear once the temperature is raised above the Curie temperature. Thus, in materials for which the Curie temperature is not too high, the size of the permanent dipole, and the accompanying lattice distortion, is temperature dependent. Thus, a small change in the temperature causes a change in the polarization within the crystal, and therefore a change occurs in the dielectric constant. This means that the electric field also changes, and the temperature change can be sensed by measuring the change in the field. The pyroelectric coefficient is defined as

$$p_{\text{pyro}} = \frac{dP}{dT}. \qquad (7.38)$$

Consider a material such as lead zirconium titanate (PZT). Here, the pyroelectric coefficient at room temperature is about 3.8×10^{-4} C/m^2K. Thus, a temperature change of 1 mK produces a change in the polarization of 3.8×10^{-7} C/m^2, a small quantity. However, the dielectric constant is 290 so that this polarization change produces a change in the electric field (in, for example, a capacitor in which the PZT is the dielectric) of 149 V/m. If the PZT film is 0.1 mm thick, the voltage change in the capacitor is 15 mV, which is easily measured. In Chapter 8, where we discuss optoelectronics, some examples of pyroelectric detectors will be presented.

7.5. MICROELECTROMECHANICAL STRUCTURES

The development of micromachining arose from the lithographic processing that has become prevalent in the semiconductor integrated circuits industry. In Chapter 6, we encountered examples of the use of oxides and photoresists to pattern specific structures in the semiconductor substrate (which was then used for local area implantation of a dopant, for example). Microelectromechanical systems (MEMS) use this technology to create novel mechanical systems or systems in which mechanical actuation is used in conjunction with electronic processing.[9] Here, the standard semiconductor processing steps of photolithography, deposition, and etching are the standard tools used to create MEMS as well. To a large extent, these systems use the microfabrication and piezoelectric properties of the materials to achieve new functionality.

Let us consider a typical sequence of events. First, processing begins with the growth (or deposition) of an "active" layer on the substrate (we will discuss a few of these later). However, this active layer is only one of the key layers. This process is shown in Fig. 7.6(a). Next, a photosensitive layer — the photoresist — is deposited on top of the structure. This is then locally exposed, either through a mask or by the use of a local energy beam (electrons, photons, or ions), as shown in Fig. 7.6(b). This results in a hole being opened in the photoresist, and the exposed area can then be etched by an appropriate liquid or gas chemical compound that attacks the deposited semiconductor, as shown in Fig. 7.6(c). Finally, we show in Fig. 7.6(d) the hole that is etched. Here, the originally deposited layer has been removed in this "hole," while the substrate is unaffected by the etchant.

An alternative approach is to use a material that is resistant to the etch as the deposited layer. This could be a separate insulator such as SiN_3 or a heavily doped Si layer (there are etches which will attack Si in general, but will not attack heavily doped material when

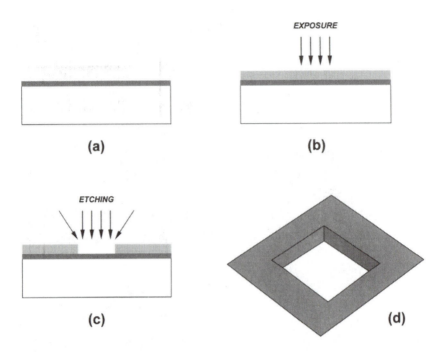

Figure 7.6. Typical processing steps for a micro-mechanical structure. (a) An active layer is formed on top of an initial substrate. (b) Photoresist is deposited on top of the active layer and is then exposed through appropriate means. (c) After exposure, a selective etch is used to remove the active layer in the region where exposure occurred, while leaving the original substrate unaffected. (d) Top view of a hole opened in the active layer through this process.

the dopant is an acceptor such as boron and the concentration approaches 10^{19} cm^{-3}). Then the etching is a two step process: (1) the deposited layer is etched into a desired shape and then (2) the semiconductor beneath the layer is removed. This is shown in Fig. 7.7. In panel (b), the SiN$_3$ layer is deposited with a built-in strain, so that it flexes upward when the underlying Si is removed. One application of this process is to make small gears and motors, some of which are shown in Fig. 7.8.

For electronic applications, the cantilever that is left after the preferential etching is used for sensing–control applications. In Fig. 7.9(a), we indicate that the cantilever, formed by preferential etching techniques, is composed of a piezoelectric material. Then a force on the cantilever is converted to an electric field through the

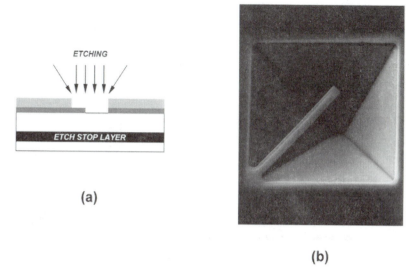

(a)

(b)

Figure 7.7. (a) An illustration of the principles of preferential etching and the use of a stop-etch layer. (b) A micrograph of a cantilever structure fabricated in silicon using preferential etching. The width of the cantilever is 10 μm and the length is 140 μm. *U. D. Vaishnav, P. R. Apte et al., "Micromechanical Components with Novel Properties," Proc. SPIE, **3321**, 287–297 (1996). Reproduced with permission.*

Figure 7.8. An example of the use of etching to form small gears. The spacer bar in this picture denotes a distance of approximately 100 μm. *Picture by H. Guckel, courtesy of University of Wisconsin.*

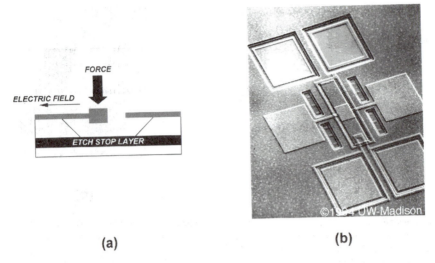

(a)   (b)

Figure 7.9. (a) Cantilever structure produced by the described selective etching process. Application of a force to the cantilever gives rise to a corresponding electric field, which is monitored to detect deflection of the beam. (b) An example of a polysilicon resonant transducer. *Picture by H. Guckel, courtesy of University of Wisconsin.*

piezoelectric effect. This field can be sensed by, for example, a transistor located near the cantilever. This is one principle of a MEMS accelerometer, such as those used in automobile air bags. One such design is shown in Fig. 7.9(b). The reverse process is to use an electric field applied to the cantilever to distort this structure and to initiate a force or deflection, as shown in Fig. 7.10(a). This reverses the sensing property discussed in the last paragraph, and is the principle behind the electronic light processor of Texas Instruments, which is used in several different brands of video projectors. Such a system is illustrated in Fig. 7.10(b).

The world of MEMS is just beginning to be fully appreciated. It is a marvelous advance which couples submicrometer-dimensioned electronic devices to micromechanical structures for the achievement of truly novel system functions. There has even been the suggestion that these techniques can put an entire chemical analysis laboratory on a single integrated circuit chip through the use of specific chemical sensors on an array of cantilevers.

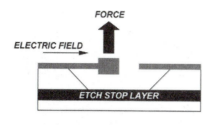

(a) **(b)**

Figure 7.10. (a) In this case, application of an electric field to the cantilever is used to produce a deflection of the cantilever beam. (b) Cantilevers of this type form the basis for electrostatic relays such as the one shown in this picture. The spacer bar in this picture denotes a distance of approximately 100 μm. *Picture by H. Guckel, courtesy of University of Wisconsin.*

REFERENCES

1. J. M. Ziman, *Principles of the Theory of Solids* (Cambridge: University Press, 1964), p. 224.
2. D. T. Paris and F. K. Hurd, *Basic Electromagnetic Theory* (New York: McGraw-Hill, 1969), p. 65.
3. D. K. Ferry, *Semiconductors* (New York, Macmillan, 1991).
4. D. Penn, Phys. Rev. **128**, 2093 (1962).
5. C. Kittel, *Introduction to Solid State Physics*, 6th ed. (New York: John Wiley, 1986), pp. 388–390.
6. E. A. Ash, *Surface Acoustic Waves for Signal Processing* (Boston: Artech House, 1989).
7. T. S. Hutchinson and D. C. Baird, *The Physics of Engineering Solids*, 2nd ed. (New York: John Wiley, 1968), pp. 482–489.
8. S. B. Lang, *Sourcebook for Pyroelectricity* (London: Gordon and Breach, 1974).
9. S. E. Lyshevsky, *Electromechanical Systems, Electric Machines, and Applied Mechatronics* (Boca Raton, Fl.: CRC Press, 2000).

PROBLEMS

1. Assume that you are given a supply of polyethylene tape and a supply of aluminum tape. The polyethylene tape is 3 in. wide, 0.0005 in. thick, and has a relative dielectric constant of 2.3. The aluminum tape is 2.5 in. wide and 0.0004 in. thick. You are asked to fabricate a capacitor of 0.1 pF capacitance in the form of a compact cylindrical roll. Describe how you might approach this problem, and give the amounts of each type of tape required for the capacitor. What is the overall diameter of the finished capacitor?

2. Two charges of $+e$ and $-e$ are separated by a distance of 1 m. What is the dipole moment? Repeat for a separation of one angstrom.

3. Consider a material with a dielectric constant of 16 and a density of 35 g/cm^3. The atoms each contribute one electron to the electron gas and have an atomic weight of 64. What is the characteristic frequency at which the electronic polarizability ceases to be important? Assume that the electronic polarizability is the main contributor to ε_r.

4. The major contribution to the dielectric constant of copper at optical frequencies is due to electronic effects. Compute the ultraviolet frequency at which the electronic polarizability begins to become ineffective.

5. A long thin cylindrical rod of electronic material has an atomic density of 3×10^{22} cm^{-3}. Each atom has a polarizability of 5×10^{-37} F/cm^2. Find the internal electric field when an axial field of 10 mV/cm is applied.

6. A capacitor is to be made from dielectric material possessing a breakdown electric field of F_{bd}. As is usual, metal plates are used for the capacitor, but a slight distortion has occurred during fabrication. As a result, one third of the area is separated from the dielectric by an air gap, whose thickness is 1 μm. The remaining two-thirds of the capacitor dielectric makes intimate contact with the metal electrodes, with a spacing of 0.5 mm. In normal operation, the dielectric has $F_{bd} \sim 20$ kV/cm, while the breakdown field for air is 30 kV/cm. The dielectric has $\varepsilon_r = 1000$. What is the effect of the air gap on (a) the capacitance and (b) the breakdown voltage of the capacitor.

CHAPTER 8

Optoelectronics

The properties of excess carriers in semiconductors is one of the more interesting aspects of this material. We first encountered excess carriers in Chapter 5, where we discussed diffusion and recombination of these carriers. They were then quite important for the operation of the $p–n$ junction diode and the bipolar transistor, which were described in Chapter 6. While the excess carriers are important in these semiconductor devices, they are also important for another class of devices, and these are the optical semiconductor devices — an area known as *optoelectronics*. Obviously, excess carriers are the result of the creation of new, nonequilibrium carriers. In the devices mentioned, these new carriers come from *injection* across the junction. They can also be generated by the absorption of an optical photon. Previously, we considered only the photoelectric effect in which the photon gave rise to a electron that was emitted from the surface. This is the case when the photon energy is larger than the work function. However, there are cases in which the photon creates an excess electron–hole pair, but these are not emitted from the surface. This photoabsorption occurs when $\hbar\omega > E_G$, where E_G is the gap between the lowest conduction band state and the highest valence band state. While these new carriers (the electron and hole created by the photon absorption) are not emitted from the surface, they are free to contribute to the conduction process through the semiconductor.

In general, the absorption of the photon, and the resulting conduction process for the electron and hole are termed *photoconductivity*. Under certain conditions, when no current is allowed to flow, an induced voltage is measured, and this is a *photovoltaic* effect. Of course, if we have excess carriers, they are free to recombine and, under the right conditions, the energy released can appear as an emitted photon. Such light emission processes can be used as a source of radiation. Examples are the light-emitting diode and the semiconductor laser.

In this chapter, we consider these optoelectronic devices and processes. We first discuss photoconductivity and photodiodes. Then we turn to the ideas of stimulated emission and population inversion. This is first discussed for a gas laser, and then the case of the semiconductor laser is treated.

8.1. PHOTODETECTION DEVICES

As already discussed, at sufficiently high frequencies, such as in the optical region, it is possible for the electromagnetic field to induce electrons from the valence band (or from an impurity level lying in the energy gap) to move into the conduction band, creating a new electron–hole pair. This is the optical absorption process. Such a process is the inverse process of recombination that was introduced in Sec. 5.7. In recombination, an electron in the conduction band combines with a hole in the valence band, and the excess energy (which is approximately the band gap energy) is lost either to heat (lattice vibrations) or to the emission of a photon. This opposite process arises from the absorption of the photon energy by an electron in the valence band, which is then excited into the conduction band, leaving a hole behind. Both of these new processes can now contribute to the conduction process. In general, there are two distinct optical absorption processes that must be considered. One occurs in materials in which the minimum of the conduction band lies at the same point (in the Brillouin zone) as the maximum of the valence band, such as GaAs. In this case, the optical transition is a vertical one, and can occur with a relatively high probability. In the other, the minimum of the conduction band does not occur at the same point (in the Brillouin zone) as the maximum of the valence band, such as Si. In this case, the transition must involve a phonon (as will be discussed later) and is, therefore, a much less likely process. In this latter case, the absorption is quite weak, but still does occur.

The important quantity in optical absorption is the absorption coefficient α, which has units of inverse length (so many absorption processes occur per unit length of the light propagation through the material). In materials such as GaAs, the process is termed a *direct* process, because the photon leads directly to the creation of an electron and a hole. The units of the momentum wave vector k are of the order of magnitude of π/a, where a is the lattice constant. This latter is of the order of 0.5 nm, which means that k is of the order of 10^8 cm^{-1}. On the other hand, the momentum wave vector of the optical wave is $E\sqrt{\varepsilon/\varepsilon_0}/\hbar c \sim 10^5$ cm^{-1}. Hence, the optical wave vector is very much smaller than the electron (or hole) wave vector. On this scale, the momentum transferred from the photon to the electron is negligible, so that the transition from the valence band to the conduction band is *vertical* (conserves momentum in the Brillouin zone). This is shown in Fig. 8.1(a) for a hypothetical band structure that supports direct transitions. In semiconductors, this

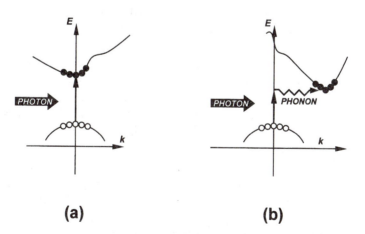

(a)   **(b)**

Figure 8.1. (a) In a direct semiconductor the top of the valence band and the bottom of the conduction band occur at the *same* value of the wavevector. (b) In an indirect semiconductor the top of the valence band and the bottom of the conduction band occur at *different* values of the wave vector.

direct transition is quite probable, since it includes only one quantum process — the absorption of the photon by the valence band electron. The absorption coefficient depends on the number of final states that are available for the process, so that it then depends on the density of states in the conduction band. As a result, the absorption coefficient can be written as

$$\alpha = \alpha_0 \left(\frac{\hbar \omega}{E_G} - 1 \right)^{1/2}. \tag{8.1}$$

For most direct gap materials, α_0 is of the order of 10^5 cm^{-1}. That is, if the photon energy $\hbar \omega$ is 10% larger than the band gap, the absorption is roughly 10^4 cm^{-1}. A specific semiconductor will vary from this rough value by a factor of 2–5, due to variations in the effective mass (of the electron and hole) and energy gap. The absorption coefficient of GaAs is shown in Fig. 8.2.

In materials where the minimum of the conduction band does not occur at the same point as the maximum of the valence band, an additional process is required. This *indirect* process occurs by the

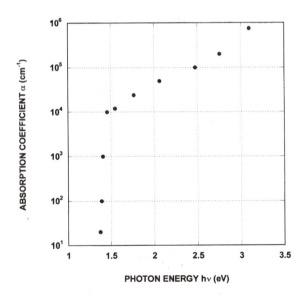

Figure 8.2. The energy dependence of the absorption coefficient of GaAs at 300 K.

absorption of a photon by an electron into a *virtual* state lying within the band gap. This virtual state has exponential decay with time, so that it must quickly interact with a lattice phonon (low energy, but large momentum transfer) to scatter into the minimum of the conduction band, as shown in Fig. 8.1(b). In computing the band structure in Chapter 4, it was always assumed that we sought states that were propagating waves. This led to forbidden regions of energy — the band gaps. In fact, this does not mean that states do not exist in these regions, only that they are not propagating waves. Instead, they decay rapidly (such as under the barrier in Sec. 3.4), so are referred to as virtual states. The electron can exist in these states for only a very short time. For this reason, the absorption process is quite weak, since it must also include two quantum processes — the interaction with the photon and the interaction with the phonon. As a result, the absorption process also has a more rapid energy dependence, and

$$\alpha_I = \alpha_{0I} \left(\frac{\hbar(\omega \pm \omega_{\text{phonon}})}{E_G} - 1 \right)^2. \tag{8.2}$$

The phonon energy is of the order of tens of meV, while the photon energy is of the order of electron volts, so the former can usually be ignored. In these indirect processes, α_{0I} is usually much smaller than α_0, and for Si is of the order of 1.5×10^3 cm^{-1}, so that the absorption in the region where the photon energy is about 10% larger than the band gap is of the order of 150 cm^{-1}. This is roughly two orders of magnitude weaker than in the direct process that occurs in GaAs. The absorption curve for Si is shown in Fig. 8.3.

8.1.1. Photoconductivity

In semiconductors at low temperatures or semiconductors in which there is only a small free-carrier concentration (due to the donors or acceptors), the number of free carriers can be increased significantly by illuminating the material with optical radiation whose photon energy is larger than the band gap. At higher temperatures, the conductivity change in this process is relatively small unless the light

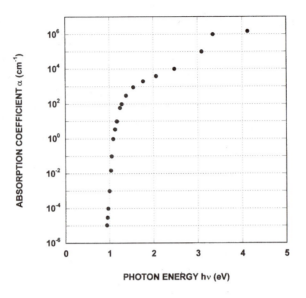

Figure 8.3. The energy dependence of the absorption coefficient of Si at 300 K.

is of very high intensity (high power per unit area of illumination). Also, at low temperatures, carriers localized on the impurity atoms can be excited into the nearby bands optically. In each of these processes, the excess carriers generated by the incident radiation leads to an increase in conductivity, a process known as photoconductivity.

If the intensity of the illumination is I_p (in watts per centimeter squared) *inside* the semiconductor (the incident radiation can be reduced by partial reflection at the semiconductor surface), the power absorbed (converted into electron–hole pairs for the fundamental absorption) is given by the *quantum efficiency* η as ηI_p. Then, the number of electron–hole pairs produced per unit length is $\eta \alpha I_p / \hbar \omega$. If we consider a thin slab of semiconductor, in which the radiation is uniform throughout the thickness (the thickness is less than $1/\alpha$), we can determine the number of excess carriers as a result of the diffusion equation, in which there is no spatial variation, as

$$\frac{d\Delta p}{dt} = G_{ex} - \frac{\Delta p}{\tau_p}. \tag{8.3}$$

Here, the generation rate, due to the external radiation, is $G_{\text{ex}} = \eta \alpha I_p / \hbar \omega$, and τ_p is the recombination time. In this process, the number of electrons created Δn is exactly equal to the number of holes created Δp. As discussed in Chapter 5, we only need to worry about the minority carriers, so have assumed in (8.3) that we are dealing with n-type material where the background free carrier density $n_0 \gg p_0$. In steady state, the excess carrier density is given by

$$\Delta p = \tau_p G_{\text{ex}} = \frac{\tau_p \alpha \eta I_p}{\hbar \omega}. \tag{8.4}$$

This excess hole concentration, the minority carrier density in n-type material, leads to a net conductivity

$$\begin{aligned} \sigma &= (n_0 + \Delta n) e \mu_e + (p_0 + \Delta p) e \mu_h \\ &\approx n_0 e \mu_e + \Delta p e (\mu_e + \mu_h). \end{aligned} \tag{8.5}$$

Here, the increase of conductivity is the photoconductivity

$$\Delta \sigma = \Delta p e (\mu_e + \mu_h). \tag{8.6}$$

This change in conductivity leads to a photocurrent, for a sample of thickness d and width w and length L,

$$i_{\text{photo}} = wd \Delta \sigma F = \frac{wd}{L} \Delta \sigma V = \frac{wd}{L} \Delta p e (\mu_e + \mu_h) V. \tag{8.7}$$

As long as the excitation is low, this leads to a photocurrent that is linear in the intensity of the radiation. The approach is useful for low light levels, but measuring small changes in the conductivity is a problematic experimental problem. In many applications, other approaches are pursued.

8.1.2. Transverse Photovoltage

In the situation where the sample is not optically thin (e.g., where $\alpha d \gg 1$), the intensity can vary throughout the thickness of the semiconductor sample. We touched on this briefly in Chapter 5, as the variation in optical illumination leads to a spatially varying excess density. That is, the number of carriers excited by the photons decreases exponentially away from the illuminated surface, as the absorption causes a decrease in the light intensity as $e^{-\alpha x}$, so that when the thickness is larger than $1/\alpha$, the illumination is no longer uniform. The spatial variation in the number of excess carriers throughout the thickness will lead to the generation of diffusion potentials through the sample, and the resulting voltage differential between the front and back surfaces is termed the *transverse photovoltage*. The illumination may be taken to occur at the top surface, and arrives normal to this surface, which is taken as the $x = 0$ plane. The sample is located in the space $x > 0$, with the slab of semiconductor oriented in the (y, z) plane with a thickness in the x direction of d. In the x direction, the photocurrent is due to the field F_x between the top and back surfaces, and

$$J_x = [e(\mu_h + \mu_e)\Delta p + n_0 e \mu_e]F_x + e(D_e - D_h)\frac{\partial \Delta p}{\partial x}. \qquad (8.8)$$

Since no terminal current flows in this direction, this current density must be set equal to zero, which causes the drift and diffusion forces to balance. This leads to the local electric field

$$F_x = -\frac{D_e - D_h}{(\mu_e + \mu_h)\Delta p + n_0\mu_e}\frac{\partial \Delta p}{\partial x} \sim -\frac{k_B T}{e}\frac{\mu_e - \mu_h}{n_0\mu_e}\frac{\partial \Delta p}{\partial x}, \qquad (8.9)$$

where we have used the Einstein relation for nondegenerate semiconductors to replace the diffusion constant by the mobility, and we have assumed that the excess density is small compared to the background electron concentration. Since the density decreases in the positive x direction, the electric field F_x points in the positive x direction if $\mu_e > \mu_h$, which is normally the case. Hence, the voltage is positive at the top surface relative to the back surface.

We can now integrate the electric field to find the voltage. In this, we introduce the mobility ratio $b = \mu_e/\mu_h$, and

$$V_{\text{TPV}} = -\int_0^d F_x \, dx = \frac{k_B T}{e n_0}\left(\frac{b-1}{b}\right)[\Delta p(0) - \Delta p(d)]. \qquad (8.10)$$

This is the voltage at the top surface relative to the back surface ($b > 1$). If the sample is electrically thick, as assumed, then we can ignore the density at the back surface, and the photovoltage is directly related to the incident light intensity through (8.4), and

$$V_{\text{TPV}} = \frac{k_B T}{e n_0} \left(\frac{b-1}{b} \right) \frac{\tau_p \alpha \eta I_p}{\hbar \omega}. \tag{8.11}$$

In Si, $b \sim 10$, so that for a doping concentration of 10^{17} cm^{-3}, a lifetime of 1 ms, a quantum efficiency of 10%, a photon energy of 2 eV (where α is about 3500 cm^{-1}) and intensity of 10 mW/cm^2, the transverse photovoltage is about 2.6 mV. If we lower the background electron concentration to 10^{15} cm^{-3}, then this value could be raised to 260 mV, but since the excess hole concentration is approximately 10^{16} cm^{-3}, the approximations are no longer valid.

8.1.3. Pyroelectric Detectors

In Chapter 7, we discussed the pyroelectric effect. In ferroelectric materials, the permanent dipole (and the large dielectric constant) disappears once the temperature is raised above the Curie temperature. Thus, in materials for which the Curie temperature is not too high, the size of the permanent dipole, and the accompanying lattice distortion, is temperature-dependent. Thus, a small change in the temperature causes a change in the polarization within the crystal, and therefore a change occurs in the dielectric constant. This means that the electric field also changes, and the temperature change can be sensed by measuring the change in the field. The pyroelectric coefficient is defined as

$$p_{\text{pyro}} = \frac{dP}{dT}. \tag{8.12}$$

Consider a material such as lead zirconium titanate (PZT). Here, the pyroelectric coefficient at room temperature is about 3.8×10^{-4} C/m^{2}K. Thus, a temperature change of 1 mK produces a change in the polarization of 3.8×10^{-7} C/m^2, a small quantity. However, the dielectric constant is 290 so that this polarization change produces a

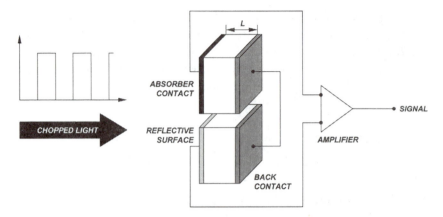

Figure 8.4. Pyroelectric detector circuit. In this circuit, the lower pyroelectric element is used to sense the environmental temperature, while the upper one is used to detect incoming thermal radiation. The net signal that is fed into the amplifier is the *difference* of these two detector signals.

change in the electric field (in, for example, a capacitor in which the PZT is the dielectric) of 149 V/m. If the PZT film is 0.1 mm thick, the voltage change in the capacitor is 15 mV, which is easily measured. In fact, this voltage is larger than the transverse photovoltage of the previous section, for reasonable parameters. Hence, this type of detector, which measures the emission temperature, rather than the light intensity, makes a good detector, especially in the infrared where most semiconductors are not sensitive.

In Fig. 8.4, we sketch a possible circuit for a pyroelectric detector. The problem is to subtract out the temperature of the environment and sense only the temperature of the preferred source (image) of the detector. This is accomplished by using *two* pyroelectric elements. One senses the incoming thermal radiation, while the second has this input blocked off, but does sense the environmental temperature. These are connected in the difference mode, so that the net signal is only the excess (above background) radiation of the incoming signal. This incoming radiation is converted to a thermal source by using an absorber on the pyroelectric material. The difference signal is sent to an amplifier. Increased sensitivity can be obtained by chopping the incoming signal, so that the difference signal periodically returns to zero, which is used as a calibration of the detector signal. Pyroelectric detectors are relatively cheap, and since they are sensitive to the infrared, an array of these detectors provides a good infrared imaging capability. These have been used in personal night vision devices and have been considered for use in automotive night vision systems. Large arrays have been used in space applications.

8.1.4. Photodiodes

It is clear from the example worked at the end of Sec. 8.1.2 that a reduction in the majority carrier concentration leads directly to a larger photovoltage. On the other hand, one cannot really reduce the majority carrier density without raising the overall resistance of the device. A solution for this is to create a photodiode. In this case, we want to absorb the photons in the depletion region of a $p-n$ junction diode. Then, the electron–hole pairs are produced in a region of low carrier density, but also in a region with a high electric field. This field will separate the electrons and holes. As shown in Fig. 8.5, the

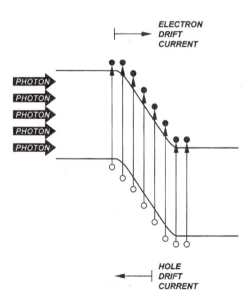

Figure 8.5. A $p-n$ junction photodiode. When the junction is illuminated with light, excess electron–hole pairs are created. A measurable voltage develops across the junction when the excess electrons and holes are swept in opposite directions across the depletion region, under the action of the net electric field that is present in this region.

electrons move from the depletion region toward the n-type region. On the other hand, the holes move toward the p-type region. If no current flows, this produces a charge induced voltage that essentially tries to forward bias the junction. Hence, a voltage is developed across the junction, and this voltage appears as an output voltage directly related to the photon flux.

In the short-current situation, where the diode works into a low-impedance load, then the excess carriers produce a photocurrent, which is in the reverse current direction. In essence, the photon creates excess carriers, which appear exactly like the excess carriers that appear at the collector junction in a bipolar transistor. These excess carriers act to increase the negative current at zero bias (or at any bias). If a large reverse bias is placed across the diode, or the diode includes an *intrinsic* region within the depletion region (a so-called p–i–n diode), and the diode is biased near to breakdown, then the excess carriers can be amplified by carrier multiplication in the reverse-biased diode.

These photodiodes can even be used as solar cells, to convert the photon flux from the sun into usable output power. However, the design is quite different. In a photodiode, where speed of response is important, one wants a small cross-sectional area of the diode, to reduce the diode capacitance, and a relatively sharp p–n junction so that the transit time is small. On the other hand, one needs to make the depletion region (or the intrinsic region) sufficiently thick that the photon flux is fully absorbed which raises the quantum efficiency of the diode. Hence, there is a trade-off on the thickness of the depletion region, but the area must remain small.

For a solar cell, on the other hand, one wants a large area to absorb the maximum number of photons from the enormous universal coverage of solar radiation. Here, however, one does not want to apply a bias voltage, as this reduces the energy conversion efficiency. The key factors for solar cells are the reverse current at zero bias and the forward voltage at zero current, both of which are induced by the photogenerated excess carriers. These two quantities together determine the maximum possible power available from the solar cell

$$P_{\max} = I_{\mathrm{SC}} V_{\mathrm{OC}}. \tag{8.13}$$

These two quantities are indicated in Fig. 8.6. The details of the diode

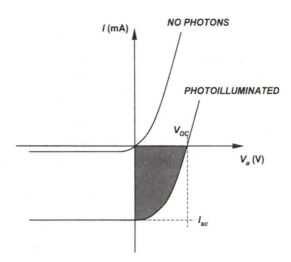

Figure 8.6. The influence of illumination on the current–voltage characteristic of a p–n junction solar cell. The maximum possible power that can be derived from the cell is determined by the quantities I_{SC} and V_{OC}, whose definitions are also indicated on this figure.

characteristics determine how much power is really available, and the ratio of the shaded area in the figure to (8.13) is termed the *fill factor*. The fill factor determines how much of the maximum available power can actually be obtained from the solar cell.

Another problem arises in both normal photodiodes and solar cells. The light comes through one of the doped regions, for example, through the n-type region, so that this region must be made quite thin and consequently must be relatively highly doped so that the depletion region forms in the oppositely doped region. This thinness is required for there to be very little absorption in this layer, as this reduces the efficiency of the device. Since the layer is thin, taking the current out from the edges presents a problem due to high resistance in the lateral directions. Hence, one wants to put contacts on the top surface, but this reduces the area available for optical absorption as the metal contact material reflects incident light. Thus, contact design to minimize the lateral resistance while maximizing the area that can absorb photons is important. This can be alleviated somewhat by fabricating a heterojunction photodiode in which the top layer is optically transparent. That is, the top doped layer, the n-type layer in this discussion, is made from a wide-band-gap material, so that photons are not absorbed in this layer, but in the narrower-band-gap p-type layer. Then, the top layer can be made sufficiently thick to reduce the lateral resistance without impacting the efficiency of the device.

Let us now consider the electrical behavior of the photodiodes. We will assume a heavily doped, but thin p-type layer at the surface and a lightly doped n-type layer below the surface, so that most of the built-in potential and the depletion width is in the n-type region. Electron–hole pairs are created in the depletion region, and the electrons drop through the built-in potential to the n-type region. The holes move to the p-type region at the surface. In general, the number of electron–hole pairs that are created at any point decreases as one moves away from the surface due to the decay of the illumination as $\exp(-\alpha x)$. Moreover, the maximum of the electric field is usually right at the interface between the two regions. Hence, the photocurrent density, which varies as $\Delta\sigma F$, would normally be at the junction interface. However, Kirchhoff's current law requires that the current density be uniform as a function of x through the junction (for a current that flows vertically through the structure). Hence, the fact that the carriers will flow slower as the field drops as one moves away from the interface leads to a buildup of the charge. One must really solve Poisson's equation, complete with the nonuniform illumination, within this region and then determine the current.

We achieve an approximation to the preceding problem by recognizing that the current at the junction interface is in a region before the charge buildup, so that we can simplify the approach by making the assumption that this region dominates the current. Then, we can say that

$$J_x \sim \Delta p e (\mu_e + \mu_h) F_{x,\max} \cdot \Delta p = \tau_p G_{ex} = \frac{\tau_p \alpha \eta I_p}{\hbar \omega}. \qquad (8.14)$$

We can take the maximum of the field from (6.17) as

$$F_{\max} = -\frac{e N_D x_n}{\varepsilon} \approx -\frac{e N_D W}{\varepsilon}, \qquad (8.15)$$

where W is the depletion width (mostly in the n-type region). Now, using (8.4) we can write the photocurrent density as

$$J_x = -\frac{\tau_p \alpha \eta I_p e \mu_h (b+1)}{\hbar \omega} \frac{eN_D W}{\varepsilon}. \tag{8.16}$$

Following the preceding example, we find that the excess density created for 2-eV photons and 10 μW/cm^2 intensity was about 10^{13} cm^{-3}. If the p-type region is doped to 10^{19} cm^{-3} and the n-type region to 10^{17} cm^{-3}, then the junction width is approximately 100 nm. Hence, the peak electric field is 160 kV/cm. If the hole mobility is 20 cm^2/Vs and $b = 10$, respectively, then the photocurrent density is 56.3 A/cm^2.

The total diode current includes the injection current, so that we can write this total current density as

$$J = J_{\text{sat}} \left[\exp\left(\frac{eV_a}{k_B T}\right) - 1 \right] - J_x. \tag{8.17}$$

We can now find the open-circuit voltage, where the total current goes to zero, to be

$$V_{\text{OC}} = V_a(J=0) = \frac{k_B T}{e} \ln\left(1 + \frac{J_x}{J_{\text{sat}}}\right). \tag{8.18}$$

Using the parameters following (6.34), with the change in parameters used here, we find that J_{sat} is approximately 2.8 nA/m^2 or 0.28 pA/cm^2. Hence, the open-circuit voltage is about 33 $k_B T$ or 0.85 V. At the same time, we can integrate the current equation (8.17) from 0 to V_{OC}, in order to find the actual power, and therefore find the fill factor. This leads to

$$P_{\text{available}} = -\int_0^{V_{\text{OC}}} J \, dV = (J_x + J_{\text{sat}})V_{\text{OC}} + J_{\text{sat}} \frac{k_B T}{e} \left[\exp\left(\frac{eV_{\text{OC}}}{k_B T}\right) - 1 \right]$$

$$= J_x (V_{\text{OC}} - k_B T/e) + J_{\text{sat}} V_{\text{OC}}, \tag{8.19}$$

where we have used (8.17) to simplify the exponential. The fill factor is then

$$ff = 1 - \frac{J_x}{J_x + J_{\text{sat}}} \frac{k_B T}{e V_{\text{OC}}} \sim 0.97. \qquad (8.20)$$

This is a very large value, and one seldom found in practice. The problems we have ignored here are the series resistance of the diode and the nonideality of the current–voltage relationship. The latter, in particular, is sensitive to traps and recombination centers that lie within the depletion region.

8.2. SPONTANEOUS AND STIMULATED EMISSION

In Sec. 3.8, we pointed out that an electron absorbed energy when it underwent a transition from an atomic level of energy E_i to one of energy E_j if $E_j > E_i$. Emission is the reverse process, in which the electron makes a transition to a final state whose energy is less than that of the initial state. If an atom is excited then the electrons are raised to higher energy levels. These excited electrons gradually decay back to their initial unexcited states. Before decaying, however, the electrons reside in the excited level for some time. The reason for this is that the transition from the upper to the lower state is a kinetic process. The wave functions corresponding to the two different energy levels are different. Therefore, in making the transition, the electron must change its wave function. The probability that this will occur is less than unity, and in some cases is exceedingly small.

It should be stressed that the electron will always eventually make its way back to the unexcited, or *ground state*. Therefore it is the transition probability per unit time that is the quantity of interest. If it is very unlikely that the transition will occur, then the transition probability per unit time is very small, and the electron will reside in the excited state, on the average, for a long time. The average length of time that the electron remains on the excited level is called the *lifetime* of that level. (When we treated electron–hole pairs earlier, this lifetime is the recombination time discussed there.) If the electron makes its way back to the unexcited—or ground—state without any outside influence and radiates its excess energy of transition,

$$\hbar\omega = E_j - E_i, \tag{8.21}$$

the process is called *spontaneous emission*, and the relative amplitude of the emission is inversely proportional to the lifetime of the electron in the upper level, E_j.

It is possible for another transition to occur in which the electron is *induced* by an outside force to make the downward transition. This effect is precisely the inverse of absorption, and is called *stimulated emission*. In absorption, the upward transition of the electron is due to its absorption of energy from the electromagnetic wave. In stimulated emission, however, the electron loses energy to the wave in the form of radiation. This extra radiation serves to reinforce or amplify the electromagnetic field.

Now why does stimulated emission—and not spontaneous emission—amplify the wave? If the transition probability per unit time for stimulated emission is proportional to the electric field, for example, this probability is a maximum when the electric field at the atom is a maximum. Since the electric field is oscillating in time (microwaves and light waves are ac signals), the transition is most likely to occur when the field is at its peak value. The field from the photon emitted by the electron during its downward transition is also at a peak at the instant of time at which it is emitted. Thus the field due to stimulated emission reinforces the electromagnetic wave.

For spontaneous emission, however, the transition probability per unit time is *not* proportional to the field of the electromagnetic wave. Therefore transitions due to spontaneous emissions produce fields which on the average are not in phase with the wave. In general, the phases of the spontaneously emitted photons are completely random and tend to cancel out. These photons produce nothing but noise in the system, and are a mechanism of energy loss.

We can formulate the relative transition probabilities of the three processes by considering thermal equilibrium processes. This was done by Einstein, and the transition probabilities involved are called the *Einstein coefficients*. Let the probability per unit time that an electron will absorb radiation be $A_{ij}\rho_\omega$, where ρ_ω is the *radiation energy density per unit frequency*, and A_{ij} is the rate of stimulated absorption (caused by the photons). Also let the probability of stimulated emission per unit time be $A_{ji}\rho_\omega$, remembering that $E_j > E_i$, and the wave frequency is related to these energies by (8.21). Finally, let the probability of spontaneous emission per unit time be B_{ji}. Then in a unit of time dt, in thermal equilibrium, the upward transitions must be equal to the downward transitions, or

$$N_j(A_{ji}\rho_\omega + B_{ji}) = N_i A_{ij}\rho_\omega, \tag{8.22}$$

where N_i and N_j are the number of electrons per unit volume in the energy states E_i, and E_j, respectively. In this treatment it is assumed that a great number of atoms are present. Equation (8.22) may be rewritten as

$$\frac{N_i}{N_j} = \frac{A_{ji}\rho_\omega + B_{ji}}{A_{ij}\rho_\omega}. \tag{8.23}$$

But, in thermal equilibrium,

$$\frac{N_i}{N_j} = \frac{f_{MB}(E_i)}{f_{MB}(E_j)} = \exp\left(\frac{E_j - E_i}{k_B T}\right) = \exp\left(\frac{\hbar\omega}{k_B T}\right), \tag{8.24}$$

where Maxwell–Boltzmann statistics have been used for a gas or for a solid for $k_B T \ll E_i, E_j$. Using this in Eq. (8.23) yields ρ_ω as

$$\rho_\omega = \frac{B_{ji}}{A_{ij}e^{\hbar\omega/k_B T} - A_{ji}}. \tag{8.25}$$

Equation (8.25) may readily be evaluated for the three coefficients in a relative fashion by comparing it with Planck's radiation law for photons, introduced in the introduction to Chapter 3. The distribution function for photons is the Bose–Einstein distribution. This distribution is

$$f_{BE}(\hbar\omega) = \frac{1}{e^{\hbar\omega/k_B T} - 1}. \tag{8.26}$$

Clearly, the relation between the coefficients is such that $A_{ji} = A_{ij}$. The ratio of B_{ji}/A_{ji} yields the coefficients that make (8.25) into Planck's radiation equation.

Recall that in Chapter 5, we found that the density of states per unit energy per unit volume was

$$\rho(E) = \frac{1}{2\pi^2}\left(\frac{2m^*}{\hbar^2}\right)^{3/2} E^{1/2}. \tag{8.27}$$

To obtain the density of modes per unit frequency, we replace E by $\hbar\omega$, using the fact that

$$\rho(\omega)\, d\omega = \rho(E)\, dE, \tag{8.28}$$

or

$$\rho(\omega) = \hbar\rho(E). \tag{8.29}$$

For the photon mass, we use Einstein's relation $E = mc^2$, so that

$$m_{\text{photon}} = \frac{E}{c^2} = \frac{\hbar\omega}{c^2}. \tag{8.30}$$

One correction must be accounted for, however. A particular energy E corresponds to two photons traveling in opposite directions. [A factor of 2 for spin was already included in (8.27), which accounts for the two polarizations of the wave. However, we must add another factor of 2 for these two photons in each polarization as well.] We can take account of this by taking half of the mass given by (8.30) for each photon. Thus the density of modes per unit volume per unit frequency is

$$\rho(\omega) = \frac{\omega^2}{\pi^2 c^3}. \tag{8.31}$$

We can now find the energy density per unit radian frequency per unit volume from

$$\begin{aligned}
\rho_\omega &= \rho(\omega)\hbar\omega f_{\text{BE}}(\omega) \\
&= \frac{\hbar\omega^3}{\pi^2 c^3}\frac{1}{e^{\hbar\omega/k_B T} - 1}.
\end{aligned} \tag{8.32}$$

This is the *Planck blackbody radiation* law. This form differs by a factor of 2π from the normal form, as we have written it in terms of radian frequency rather than frequency. Converting to the latter adds the factor of 2π through the relation (8.28). Comparison of this last result with (8.25) leads to

$$A_{ji} = A_{ij}, \qquad \frac{B_{ji}}{A_{ji}} = \frac{\hbar\omega^3}{\pi^2 c^3}. \qquad (8.33)$$

From the first equation of (8.33), it is apparent that absorption and stimulated emission are inverse processes, and that, for these two processes, the probability of a downward transition is equal to that of an upward transition. Spontaneous emission has a low probability compared with induced emission, except at extremely short wavelengths, so that if the upper level has a reasonably long lifetime, spontaneous emission is not an important process. Since the probabilities of upward and downward stimulated transition are equal, it is apparent that for emission to exceed absorption, $N_j > N_i$. This is contrary to the normal densities at thermal equilibrium, and leads to *population inversion*.

The results of (8.33) raise an interesting question. If a system is to amplify a signal with a radian frequency ω, then the system must emit more energy of units $\hbar\omega$ than it absorbs. Equation (8.33) says that stimulated emission and absorption are reciprocal processes, since $A_{ji} = A_{ij}$ Hence the probability of a downward transition, or emission, is exactly equal to that of an upward transition, or absorption. The total number of absorbing transitions per unit time is the upward transition probability times the number of electrons in the lower or initial state, N_i. Thus the absorption is proportional to

$$N_i A_{ij} \rho_\omega. \qquad (8.34)$$

The emission probability per unit time is the downward transition probability times the number of particles in the upper state, N_j. Thus the stimulated emission is proportional to

$$N_j A_{ji} \rho_\omega. \qquad (8.35)$$

Using (8.33), we find that the net emission is the difference between the stimulated emission and the absorption, as

$$(N_j - N_i)A_{ji}\rho_\omega. \tag{8.36}$$

Therefore, if a net amplification is to result from the system, there must be more particles in the upper state, or

$$N_j > N_i. \tag{8.37}$$

But, in thermal equilibrium, (8.24) tells us that

$$N_{j,\text{eq}} = N_{i,\text{eq}}e^{-\hbar\omega/k_B T}. \tag{8.38}$$

Thus, in thermal equilibrium, $N_j < N_i$, since N_i is at the lower energy level. This means that the medium is normally non-amplifying and that the absorption of energy is greater than the stimulated emission of energy.

If one could achieve a negative temperature, however, (8.38) would admit a case in which $N_j > N_i$. It is possible to drive the system out of thermal equilibrium, so that a condition of (8.37) can be met. However, such a condition is a nonequilibrium one, and neither Fermi–Dirac nor Boltzmann statistics are strictly appropriate. Thus, Eq. (8.38) is not the correct equation to use. Such a state — in which the normal population relationships of $N_i > N_j$ are inverted — is often called a *negative temperature state*, in analogy with (8.38). The condition in which $N_j > N_i$ is one in which the normal density ratio of the two energy levels has been reversed or inverted. For this reason, the nonequilibrium condition, in which (8.37) is satisfied, is called a condition of *population inversion*. In such a state, stimulated emission can be greater than absorption, and amplification can occur. A medium in which this happens is often called an *active medium*.

8.3. LASERS

It is exceedingly difficult to obtain population inversion in a two-level system. In a three-level system, however, in which two of the levels differ by the signal energy $\hbar\omega$, it is considerably easier to obtain pop-

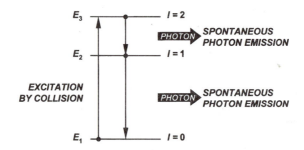

Figure 8.7. A three-level system that may be used to realize a laser.

ulation inversion between the two levels of interest. Such a system is depicted in Fig. 8.7. In thermal equilibrium, all N $(= N_1 + N_2 + N_3)$ electrons reside in level 1. If an excitation, such as a direct-current discharge, is applied, a portion of the electrons are excited to level 3. The excitation is chosen specifically so that the excited electrons from level 1 go only to level 3. The upward excitation, or absorption, transition rate is

$$W_{13} = A_{13}\rho_{exc}, \tag{8.39}$$

where ρ_{exc} is the *exciting energy density per unit frequency*. In addition, there is stimulated emission from level 3 to level 1, given by the rate

$$W_{31} = A_{31}\rho_{exc}, \tag{8.40}$$

and spontaneous emission from level 3 to level 1, given by the rate B_{31}. In the preceding, $A_{13} = A_{31}$; hence $W_{13} = W_{31}$. It is also possible for electrons that have been excited to level 3 to decay to level 2. Since this is a downward transition without any external radiation at the proper frequency (it is usually a nonradiative decay involving collisions of the electrons or atomic energy), only the spontaneous emission rate B_{32}, is considered. The signal to be amplified is chosen to have a frequency corresponding to

$$\omega = \frac{E_2 - E_1}{\hbar}, \tag{8.41}$$

so that it corresponds to the $2 \rightarrow 1$ transition. Because of the presence of this signal, there is absorption at a rate

$$W_{12} = A_{12}\rho_\omega \tag{8.42}$$

and stimulated emission at a rate

$$W_{21} = A_{21}\rho_\omega. \tag{8.43}$$

In addition, there is spontaneous emission from level 2 to level 1 at a rate B_{21}.

By applying the excitation, we can achieve a nonequilibrium situation for the three-level system. We wish to achieve a population inversion between levels 2 and 1 such that $N_2 > N_1$. To investigate this, we can write the rate of change of the population of the upper two levels as

$$\frac{dN_3}{dt} = W_{13}N_1 - (W_{31} + B_{31} + B_{32})N_3, \tag{8.44}$$

$$\frac{dN_2}{dt} = B_{32}N_3 - (W_{21} + B_{21})N_2 + W_{12}N_1.$$

A third requirement is that the total number of electrons be constant, or

$$N = N_0 = N_1 + N_2 + N_3. \tag{8.45}$$

In steady-state operation, the number of electrons in each level is constant in time, so that the derivative terms are zero. Then we can rewrite (8.44) and (8.45) in matrix form as

$$\begin{bmatrix} W_{13} & 0 & -(W_{31} + B_{31} + B_{32}) \\ W_{12} & -(W_{21} + B_{21}) & B_{32} \\ 1 & 1 & 1 \end{bmatrix} \begin{bmatrix} N_1 \\ N_2 \\ N_3 \end{bmatrix} = \begin{bmatrix} 0 \\ 0 \\ N_0 \end{bmatrix}. \tag{8.46}$$

Since only the two populations N_1 and N_2 are of interest, their ratio can be found as

$$\frac{N_2}{N_1} = \frac{W_{13}B_{32} + W_{12}(W_{31} + B_{31} + B_{32})}{(W_{21} + B_{21})(W_{31} + B_{31} + B_{32})}. \tag{8.47}$$

In the discussion of the Einstein coefficients, we pointed out that the spontaneous emission rate is generally small compared with a strong stimulated emission rate. In this case, one can neglect the B_{31} term, since it is small compared with W_{31}. Generally the material is also chosen so that the transition from $3 \rightarrow 2$ is very fast, or $B_{32} \gg W_{31}$. Thus, when we use the facts that $W_{12} = W_{21}$ and $W_{13} = W_{31}$, (8.47) becomes

$$\frac{N_2}{N_1} = \frac{W_{21} + W_{13}}{W_{21} + B_{21}}. \tag{8.48}$$

It is sometimes more valuable to write (8.48) in terms of the percentage of population inversion, or the net downward transition rate as a function of the total possible rate. We can do this by writing

$$\frac{N_2 - N_1}{N_0} = \frac{W_{13} - B_{21}}{W_{13} + B_{21} + 2W_{21}}. \tag{8.49}$$

For strong population inversion, (8.49) should be near its maximum value of 1. Thus, W_{13} must be made large, and a strong exciting source is required. Population inversion will occur if $W_{13} > B_{21}$ for the present case, in which all the excited electrons decay from level 3 to level 2 rather than back to level 1 ($B_{32} \gg W_{31}$). The total number of electrons N_0 depends on the thermal equilibrium density of the electrons in the unexcited level 1 (ground state) and hence on the density of atoms in the medium.

To create a laser, we now have to put this active medium in a resonator, or cavity. In the optical laser, this cavity is formed by two mirrors which reflect the emitted light from one mirror back through the gas to the second mirror. In this way, the light is amplified by stimulating transitions with the proper phase coherency. An optical device with these two mirrors, known as a *Fabry–Pérot resonator*, is used to form the resonating structure. Any light within the structure—light that is propagating normal to one of the mirrors—is trapped within the structure as it bounces back and forth between the mirrors. Such an arrangement provides a high-Q resonant structure, provided that the distance between the mirrors is given by

$$L = \frac{n\lambda}{2},\qquad\qquad(8.50)$$

where n is an integer, which is the same condition as our quantum well in Chapter 3. In essence, we make an optical quantum well for the radiation waves.

In the gas laser, the active medium is an ionized gas. This gas is ionized, and pumped, by a discharge, either a high-frequency rf discharge or a dc discharge. The most common gas laser is the He–Ne laser, in which the neon atoms provide the electron energy levels involved in the transitions. The active gas is contained in a glass tube located in the Fabry–Pérot cavity, as shown in Fig. 8.8. The He–Ne laser is primarily a four-level, rather than a three-level, laser, but the general approach is the same as that given above for the three-level maser. The glass windows at the end of the laser are canted at an angle to the direction of propagation. This angle, called the *Brewster angle*, makes it possible for the beam to pass through the windows unattenuated. The two most commonly used laser transitions for the He–Ne system are at a wavelength of 632.8 nm, and at a wavelength of 3.39 μm, or 3390 nm, although there are many tens of active laser lines in the gas. One other common one lies between the two just mentioned, at a wavelength of 1.15 μm.

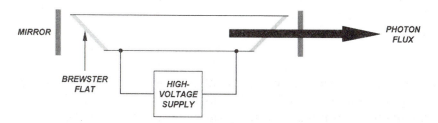

Figure 8.8. The key components of a gas laser are shown.

Laser action has been observed in most of the elements of the periodic table. The He–Ne laser discussed here was one of the earliest, and perhaps still the most common. Other important solid state lasers include those formed out of neodymium-doped glass (Nd^{3+} atoms in a host lattice made of silicate or borate glass, for example). Important gas lasers include argon, krypton, carbon dioxide, and nitrogen.

8.4. SEMICONDUCTOR LASERS

The semiconductor injection laser differs from the gas lasers in that the principle of operation does not depend on an external source or pump of exciting rf frequency. The injection laser is basically a two-level laser and is driven by the diode current itself. The active mechanism is the recombination of electrons and holes injected across the junction in a $p-n$ diode. In most semiconductors with indirect band gaps, this recombination energy primarily produces *phonons*, or heat vibrations of the lattice, because the indirect gap is not conducive to the emission of photons. In direct-band-gap semiconductors, such as GaAs, on the other hand, the recombination may take the form of optical radiation, and is a highly probable process. A *direct band gap* is one in which the minimum of the conduction band and the maximum of the valence band occur at the same point in k-space, i.e., they correspond to the same value of the crystal momentum, as was shown in Fig. 8.1. Here, the recombination process is the inverse of the generation process shown in the figure.

In germanium, for instance, the conduction band minima are located at the edge of the Brillouin zone along the (111) axes, while the valence band maximum is at the zone center (000). Thus germanium is an indirect-gap semiconductor, and is not likely to produce lasing action. In silicon, the conduction band minima are along the (100) directions, while the valence band maximum is at the zone center (000). Hence, silicon is also an indirect gap semiconductor and is not likely to produce lasing action. In GaAs, on the other hand, both the minimum of the conduction band and the maximum of the valence band are located at the center of the Brillouin zone, so that lasing action is possible.

To illustrate the manner in which the laser operates, consider a $p-n$ junction in which both sides are doped to degeneracy (the Fermi energy level lies either in the band or very near to the edge of it, preferably the former). This gives a very sharp cutoff of the Fermi–Dirac distribution. The diode is forward biased, so that electrons are injected from the n-region into the p-region. It is of interest to consider an electron energy $E(k_i)$ in the conduction band and $E(k_j)$ in the valence band. In thermal equilibrium, the number of electrons per unit volume per unit energy range is given, by analogy to (5.30) without the integration of this latter equation, by

$$n(E_i) = \rho_c(E_i) f_{\text{FD}}(E_i) = \frac{\rho_c(E_i)}{1 + \exp\left[\dfrac{E_i - E_F}{k_B T}\right]} \qquad (8.51)$$

where $\rho_c(E_i)$ is the density of states (5.31). Similarly, the number of holes at energy E_j is given by

$$p(E_j) = \rho_v(E_j)[1 - f_{\text{FD}}(E_j)] = \frac{\rho_v(E_j)}{1 + \exp\left[-\dfrac{E_j - E_F}{k_B T}\right]}. \qquad (8.52)$$

When the diode is forward biased, the number of electrons on the p-type side of the junction is increased, and the number of holes on the n-type side of the junction is also increased. The Fermi energy on the two sides of the junction differ by the applied bias eV_a. In Maxwell–Boltzmann statistics, we can then relate the minority density to the equilibrium by this voltage. With the Fermi–Dirac statistics, this is more difficult. Instead, we introduce a *quasi-Fermi level* for electrons E_{Fe} and for holes E_{Fh} so that the minority carrier concentration in the p-type region can be written in analogy with (8.51) as

$$n_p(E_i) = \frac{\rho_c(E_i)}{1 + \exp\left[\dfrac{E_i - E_{\text{Fe}}}{k_B T}\right]}. \qquad (8.53)$$

Here, $E_{\text{Fe}} - E_F = eV_a$, where the Fermi level is evaluated on the p-type side. Similarly, we mimic (8.51) to define the number of holes on the n-type side as

$$p_n(E_j) = \frac{\rho_v(E_j)}{1 + \exp\left[-\dfrac{E_j - E_{\mathrm{Fh}}}{k_B T}\right]}. \tag{8.54}$$

The absorption of the electromagnetic radiation, in the junction region where both types of excess carrier are present, is proportional to the number of carriers that are able to make the transition across the energy gap. Hence, the probability for absorption is proportional to the number of electrons (full states) in the valence band and the number of holes (empty states) in the conduction band. We can write this as

$$N_{\mathrm{abs}} = W_{\mathrm{vc}} \frac{\rho_v(E_j)}{1 + \exp\left[\dfrac{E_j - E_{\mathrm{Fp}}}{k_B T}\right]} \frac{\rho_c(E_i)}{1 + \exp\left[-\dfrac{E_i - E_{\mathrm{Fe}}}{k_B T}\right]}. \tag{8.55}$$

Stimulated emission is proportional to the number of electrons in the conduction band and the number of holes in the valence band, so that

$$N_{\mathrm{em}} = W_{\mathrm{cv}} \frac{\rho_v(E_j)}{1 + \exp\left[-\dfrac{E_j - E_{\mathrm{Fp}}}{k_B T}\right]} \frac{\rho_c(E_i)}{1 + \exp\left[\dfrac{E_i - E_{\mathrm{Fe}}}{k_B T}\right]}. \tag{8.56}$$

The two transition probabilities W_{vc} and W_{cv} are equal, due to the same arguments that led to the equality of A_{ij} and A_{ji} (these are the equivalent parameters in the junction). The inclusion of both the Fermi–Dirac distribution and the complement $(1 - f_{\mathrm{FD}})$ is a result of the Pauli exclusion principle. So that emission processes dominate absorption processes, we require that

$$N_{\mathrm{em}} > N_{\mathrm{abs}}. \tag{8.57}$$

If we now use (8.55) and (8.56), and cancel all the common terms, we are left with the requirement that

$$\exp\left[\frac{E_j - E_{\mathrm{Fp}}}{k_B T}\right] > \exp\left[\frac{E_i - E_{\mathrm{Fe}}}{k_B T}\right]. \tag{8.58}$$

The conservation of crystal momentum ensures that $k_i = k_j$, and

$$E(k_i) - E(k_j) = \hbar\omega > E_G, \qquad (8.59)$$

where the latter inequality arises from the fact that the energy levels must lie in the bands. Thus, we may rewrite inequality (8.58) as

$$E_{Fe} - E_{Fp} > \hbar\omega > E_G. \qquad (8.60)$$

Equation (8.60) is essentially the requirement of population inversion. Since $\hbar\omega > E_G$, it is apparent that for stimulated emission and laser action to occur, the level of injection of electrons into the p-region must be high enough so that it would correspond to degenerate n-type material if the holes were not already present. The requirement for population inversion is that the level of injection of electrons into the p-region is so high that the valence band is almost completely depleted of electrons at energies near the equilibrium Fermi level. The requirement of (8.60) with respect to the quasi-Fermi levels is equivalent to requiring that both sides of the p–n junction be doped to degeneracy. This situation is shown in Fig. 8.9.

The injection diode lasers are normally made so that the light propagates in the plane of the junction, and a Fabry–Pérot resonator is made by polishing two faces of the diode perpendicular to the plane of the junction. The polishing produces a high degree of reflection at these surfaces, because of the dielectric discontinuity. For normal incidence, the reflection coefficient at the GaAs–air interface

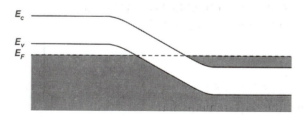

Figure 8.9. A semiconductor laser can be realized using a degenerately doped p–n junction. The band-bending diagram is shown at thermal equilibrium in this figure. Note how the Fermi level lies in the valence and conduction bands on opposite sides of the junction.

is given by

$$R = \frac{\sqrt{\varepsilon_s} - \sqrt{\varepsilon_0}}{\sqrt{\varepsilon_s} - \sqrt{\varepsilon_0}} \sim 0.54. \qquad (8.61)$$

While this reflection coefficient is not all that high, it is sufficient for the purpose. GaAs diodes emit radiation at about 840 nm, which is in the near-infrared range of the spectrum. InP, GaP, InAs, and many other direct-gap semiconductors are also used to make injection lasers. Since the diode itself forms both the active medium and the resonator, the only losses are those internal to the diode and the losses due to transmission. The population inversion must be large enough for the gain of the laser to exceed the losses in the cavity. This can be aided by using heterojunction lasers, in which the p–n junction is a material with desired wavelength, but which is embedded between two materials with larger band gaps. This confines the excess carriers in the quantum well region (the junction region) and prevents their diffusion away from the junction. Such a structure enhances the recombination and produces better lasers, with lower threshold currents. (The heterojunction laser resulted in the Nobel prize for Krömer and Alferov in 2000.) The quantum well can also push the laser transition to higher energy due to the confinement quantization of the electron and hole states. This can move the light emitted into the visible region. For fiber communications applications, the desired wavelength is about 1.5 μm (~ 0.8 eV). For this application, InGaAs is the preferred material for the quantum well junction region, as its band gap is a good match with almost equal concentrations of InAs and GaAs in the alloy.

PROBLEMS

1. A particular semiconductor material has a background donor doping of 10^{13} cm^{-3} and an electron mobility of 3800 cm^2/Vs, with $b = 2.5$. The semiconductor is illuminated over its surface with a 1 mW/cm^2 laser beam (inside the semiconductor). If the semiconductor material is 0.1 cm thick, $\alpha\eta = 10^4$ cm^{-1}, and $\tau_p = 1$ ms, calculate the ratio of the photocurrent to the dark-current

density for an applied field of 5 V/cm. The current is taken from the top and back surfaces and the voltage is applied to these surfaces.

2. For the same parameters as Prob. 1, assume that no current is allowed to flow and compute the transverse photovoltage.

3. Calculate (a) the ratio of the Einstein coefficients A/B and (b) the ratio of spontaneous to stimulated transitions for a wavelength of 693 nm at room temperature. Assume the index of refraction is 1.

4. A maser is a microwave version of the laser. In constructing one of these, a set of atomic transitions are found with relative energy levels of $E_1 = 0$ eV, $E_2 = 50$ μeV, and $E_3 = 75$ μeV. If these three levels are to be used for maser action, what should the excitation (pump) frequency and active (stimulated emission) frequency be?

5. What causes the image of a laser beam on a piece of paper to appear as if it consisted of a large number of bright points? Why do these points appear to change brightness as your eye is moved?

6. The band gap of GaAs is 1.4 eV, and the electron and hole effective masses are $0.067m_0$ and $0.48m_0$, respectively. What minimum donor and acceptor concentrations are required to make lasing action possible in a GaAs p–n junction?

CHAPTER 9

Magnetic Materials

Magnetic materials have been a keystone material in a variety of applications found in engineering. These include motors, generators, and memories for computers. Indeed, in the earliest computers, magnetic materials were used as the active medium in large rotating drums, where the information could be found by sensing heads. Later, these would be replaced by magnetic cores for random access memory and by disks for archival memory. While semiconductors devices have replaced the core memories in random access memories, magnetic disks remain the central archival memory on nearly all computers. One reason for this is their low cost. While semiconductor memory has increased in density by roughly a factor of 4 every 3 years, so too has the density of magnetic memory increased. Indeed, since the cost per bit of magnetic memory has remained almost a factor of 10 cheaper than semiconductor memory during this entire "scaling period," it is unlikely that this application will disappear any time soon.

Most types of solids exhibit some magnetic effects when placed in a magnetic field. The effects can occur due to many different mechanisms. It is useful to classify the effects which may occur, and the types of solids which exhibit them. However, since one solid may exhibit several phenomena, it is more feasible to discuss the effects themselves and the manner in which they occur. The quantity that

best characterizes the magnetic properties is the magnetic susceptibility χ_m, which is the magnetic analog of the electric susceptibility
discussed above. This susceptibility relates the magnetic flux density
B, the magnetic field intensity H, and the magnetic moment μ_m.

Before we continue to discuss magnetic effects as they occur in
solids, let us pause to review the concepts of magnetic flux density,
magnetic field intensity, and magnetic moment. The magnetic flux
density is precisely what its name implies. It is the density of lines of
flux, and is measured in Tesla (or Weber per square meter in older
units), as

$$B = \frac{\Phi}{A},$$
(9.1)

where A is the area of interest and Φ is the total flux passing through
the area. In electrical circuits, the flux density is important for its
effect in producing magnetic forces and fields. We found this in Sec.
5.5, where we introduced the Lorentz force and studied electron
transport in a magnetic field. The magnetic field intensity is a
measure of the flux intensity and is related to the current producing
the flux by the fact that the integration of H around a closed path is
equal to the *total* current enclosed. Thus, by summing over the path
length, we obtain a relation between the magnetic field intensity and
the current. The integral can be written as

$$\oint \mathbf{H} \cdot \mathbf{dl} = I = \iint_A \mathbf{J} \cdot \mathbf{n} \, dA.$$
(9.2)

Here, the contour integration in the first term encloses the area A,
with normal vector $\mathbf{n}$, through which the current density passes. The
direction of integration is taken in the *right-hand* sense, so that a
right-hand screw progresses in the direction of I as the screw rotates
with dl. Moreover, the magnetic field intensity is related to the
magnetic flux density by a basic parameter of the medium. The
relationship between the two is

$$B = \mu_r \mu_0 H,$$
(9.3)

where μ_0 is *the permeability of free space*, given by $4\pi \times 10^{-7}$ Wb/(m-A), and μ_r is the *relative permeability*. As with the relative dielectric constant, μ_r is a heuristic constant that is used to simplify the equations in such a manner that the internal fields can be included and treated merely by modifying μ_0. A magnetic dipole is, very simply, a current loop. *A* magnetic dipole moment μ_m is produced by a current flowing through the loop with enclosed area *A* such that

$$\mu_m = \oint \mathbf{I} \times \mathbf{dl} = \mathbf{n}IA, \qquad (9.4)$$

where **n** is a normal vector to the plane of the area enclosed by the contour integral. The magnetic dipole moment μ_m has a direction of the ever-present right-hand screw when rotated in the direction of the current *I*. If there are many small magnetic dipoles, we can speak of the magnetic dipole moment per unit volume, or *magnetization per unit volume*, which is symbolized by *M*.

Let us consider an example with a current of 1 mA and a defined circular area of 1 cm^2. This circular area has a radius of 0.56 cm and a circumference of 3.54 cm. If the current flows through this area, then the current density is 1 mA/cm^2, and gives rise to a magnetic field intensity $H = 0.28$ mA/cm $= 0.028$ A/m, directed azimuthally around the current (around the circumference of the area). Now, consider this magnetic field intensity directed through the same size area (now the field intensity is directed normal to the plane of the area). Using the free space permeability μ_0, the flux density *B* is 3.54×10^{-8} T, and the flux flowing through the area is 3.54×10^{-12} Wb. If the relative permeability were 10^4, then we would increase *B* to 3.54×10^{-4} T, and the flux flowing through the area would be 3.54×10^{-8} Wb.

9.1. MAGNETIC SUSCEPTIBILITY

Consider a cylinder of material of permeability $\mu = \mu_r\mu_0$. In a magnetic material such as this, there will be induced magnetic dipoles, just as there were induced electric dipoles in the dielectric media of Chapter 7. These induced dipoles contribute additional magnetic fields to the external fields, so that the fields within the material are different from those outside the material. It is to be expected that a magnetic susceptibility $\chi_m = \mu_r - 1$, which relates the magnetization M to the field intensity H, will occur. The question is just how we determine this quantity. Here, we present a *gedanken* experiment to explain how the magnetization may be evaluated. In the cylinder of material, the field intensity H is that of the applied external field, and B is given by

$$B = \mu_r \mu_0 H. \tag{9.5}$$

Now, consider a large cylinder of magnetic material, to which a magnetic field intensity is applied along the cylindrical axis. This is shown in Fig. 9.1. Within this cylinder, the magnetic flux density B is given by (9.5). Let us assume that, from the inside of the cylinder, we remove a smaller cylinder of length dl and cross-sectional area dA, so that $dV = dl\,dA$ (Fig. 9.1). When this cylinder is removed, the magnetic induction, or flux density B, in this small volume will change. However, let us suppose that a small current I' flows around the walls of this small cavity (Fig. 9.2). This current is made just large enough in magnitude that the magnetization produced is sufficient to make B_c in the cavity the same as B in the main cylinder. That is, the extra current is made sufficiently large to raise the flux density in the

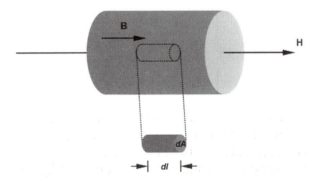

Figure 9.1. *Gedanken* experiment in which the magnetization of a magnetic cylinder is determined by removing a small volume from the inside of the cylinder.

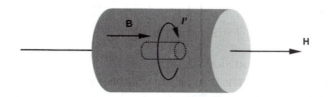

Figure 9.2. In the *gedanken* experiment of Fig. 9.1 a small current will flow around the surface of the cavity inside the cylinder in the presence of an external magnetic field.

cavity to the level within the remaining solid block. Then,

$$B_c = B. \tag{9.6}$$

Now, the small volume that was removed must have had the same magnetization as that produced by the current I'. Inside the small cavity $\mu_r = 1$, and

$$B_c = \mu_0 H_{\text{cavity}}. \tag{9.7}$$

On the other hand, inside the cylinder, we have

$$B = \mu_r \mu_0 H. \tag{9.8}$$

Here, the total H includes the role of the current generating the magnetization. Comparing (9.6) through (9.8) shows us that

$$\mu_0 H_{\text{cavity}} = \mu_r \mu_0 H \tag{9.9}$$

or

$$H_{\text{cavity}} = \mu_r H. \tag{9.10}$$

The change in magnetic field in the volume of the small cavity is $H_{\text{cavity}} - H$, and this must be produced by the magnetization of the material removed. This can be shown by calculating the current I'. The current necessary to build up B_{cavity} is (from Fig. 9.2)

$$I' = H_{\text{cavity}}\, dl - H\, dl \tag{9.11}$$

or, using (9.10),

$$I' = (\mu_r - 1)H\,dl. \qquad (9.12)$$

This current loop produces a magnetic dipole moment

$$\mu_m = I'\,dA = (\mu_r - 1)H\,dV. \qquad (9.13)$$

and the magnetization per unit volume is just

$$M = (\mu_r - 1)H = \chi_m H. \qquad (9.14)$$

The last expression defines the magnetic susceptibility χ_m as

$$\chi_m = \mu_r - 1. \qquad (9.15)$$

There are several types of magnetism observed in solids. The three most important types are *diamagnetism*, in which $\chi_m < 0$, *paramagnetism*, in which $\chi_m > 0$ and small, and *ferromagnetism*, observed in iron-type materials, in which $\chi_m > 0$ and very large. In the remainder of this chapter, we shall discuss each of these three in some detail and give typical origins. We will also discuss some novel new applications.

9.2. DIAMAGNETISM

An external magnetic field always produces some diamagnetism in atoms and ions, since the induced magnetic moments in these atoms and ions is such as to tend to oppose any magnetic field buildup. This is the microscopic, or atomic, analog of *Lenz's law*. When the magnetic field is changed, it induces currents which are of such a direction and nature that the additional fields produced tend to oppose the original change of the magnetic field. Generally this effect is small, and other magnetic effects dominate. However, in certain cases, the diamagnetism is observable. As an example of diamagnetism, consider the susceptibility for the case of a single electron orbiting a nucleus, with an orbit radius of r. The frequency of revolution of the electron is ω_0. The magnetic moment of the electron is just

$$\mu_m = IA = \left(-\frac{e\omega_0}{2\pi}\right)(\pi r^2), \qquad (9.16)$$

where $\omega_0/2\pi$ is the frequency of revolution of the electron and $-e\omega_0/2\pi$ is the current past a point. Then

$$\mu_m = -\tfrac{1}{2}er^2\omega_0. \qquad (9.17)$$

In general, if there is an ensemble of atoms such as these, the magnetic moments are randomly oriented and tend to cancel one another. If the orbits are rigidly oriented so that the magnetic moments cannot align themselves along the external magnetic field, then the only contribution to the total magnetization comes from the modification of μ_m by the external field.

However, consider the single atom and let the magnetic flux density slowly increase from 0 to B, so that the orbit encloses a flux Φ. As the flux increases, a voltage is induced in the orbit, so that the electron sees an electric field from

$$V = -\int E\, dl = \frac{d\Phi}{dt}, \qquad (9.18)$$

or

$$2\pi r E = -\frac{d\Phi}{dt}. \qquad (9.19)$$

Here, we have used Faraday's law, and assumed that E is directed along the circumference of the orbit. Since

$$B = \frac{\Phi}{\pi r^2}, \qquad (9.20)$$

it follows that

$$E = -\frac{r}{2}\frac{dB}{dt}. \qquad (9.21)$$

Because E is produced while B is increasing, an additional force $-eE$ acts on the electron during this time. We know from Lenz's law that the induced fields act in such a manner as to oppose any change of B, so that this should produce a diamagnetic effect. We can obtain the change in angular momentum of the electron under this new force from the torque acting on the electron. In time dt, the change of angular momentum appears as a change of frequency $d\omega$, so that

$$\Delta \mathbf{L} = \mathbf{r} \times \mathbf{F} \, dt, \tag{9.22}$$

and, inserting the value of the angular momentum,

$$mr^2 \, d\omega = rF \, dt = -eEr \, dt = \frac{er^2}{2} \, dB. \tag{9.23}$$

Integrating the right-hand side from 0 to B, and the left-hand side from ω_0 to ω gives

$$m \int_{\omega_0}^{\omega} d\omega = \frac{e}{2} \int_0^B dB, \tag{9.24}$$

so that

$$\omega = \omega_0 + \frac{eB}{2m}. \tag{9.25}$$

Equation (9.25) says that the angular frequency of the orbit is changed by an amount $eB/2m$. This $\Delta\omega$ is called the *Larmor frequency* of the electron

$$\omega_L = \frac{eB}{2m}. \tag{9.26}$$

We note in particular that this is just one-half of the cyclotron frequency that was introduced in Chapter 5. Here, we deal with a change in frequency rather than an orbiting frequency. This change in frequency produces a change in the loop current and hence in the magnetic moment. The induced moment is given as

$$\mu_{m,\text{ind}} = -\frac{er^2}{2} (\Delta\omega) = -\frac{1}{4}\left(\frac{er^2}{m}\right) B, \tag{9.27}$$

where we have used (9.17) for μ_m. This induced magnetic moment is in a direction opposite to that of the applied magnetic field. Thus, the susceptibility is negative and the effect is a diamagnetic one. If there are n atoms per unit volume, each with Z electrons, then the total magnetic susceptibility per unit volume is

$$\chi_m = -nZ\left(\frac{e^2\bar{r}^2}{4m}\right)\mu_0,\qquad(9.28)$$

where $\bar{r}^2$ is a suitable average over r due to the fact that the magnetic induction B makes different angles with the orbits of various electrons. In the simplest case, $\bar{r}^2 = 2r^2/3$.

9.3. PARAMAGNETISM

If the magnetic dipoles induced by the orbital angular momentum are free enough that they can align themselves along the magnetic induction B, then the fields produced by these dipoles aid the external field and the medium is termed *paramagnetic*. These permanent magnetic moments may arise from the orbital angular momentum, as discussed above. They can also occur due to the spin angular momentum of the electrons as well as the nuclear spin of the total atom.

The orbital angular momentum of the electron is $mr^2\omega$, which is denoted by L. Then, from (9.27), we can write

$$\mu_{m,\text{induced}} = -\frac{e}{2m}L.\qquad(9.29)$$

In the Bohr atom, it was observed that $L = n\hbar$, where n is an integer. A more exact solution of the Schrödinger equation for the hydrogen atom (see Appendix A) shows that the Bohr postulate on the angular momentum was inaccurate, and must be replaced by

$$L = \sqrt{l(l+1)}\,\hbar,\qquad(9.30)$$

where l is an integer related to the orbital angular momentum in the azimuthal direction. Thus,

$$\mu_m = \left(-\frac{e\hbar}{2m}\right)\sqrt{l(l+1)}. \qquad (9.31)$$

The quantity $(-e\hbar/2m)$ is called the *Bohr magneton*, β, and has the value 9.27×10^{-24} A-m^2. The value of l depends on the electron structure of the atom. Thus the total magnetic moment for an atom will be a sum of the individual electron contributions and must be made by summing over the values of l for each electron.

Since the spin is a simpler angular momentum system, we will discuss this aspect first. The momentum can be written simply as

$$L_s = s\hbar, \qquad s = \pm\tfrac{1}{2}, \qquad (9.32)$$

although experimentally, it is found that an additional term, g_s, the *Landé splitting factor*, a fudge factor, must be included, and

$$L_s = g_s s\hbar. \qquad (9.33)$$

The factor g_s that must be included is a reflection of the fact that there is a strong coupling between the orbital and spin angular moments. The spin angular momentum reflects the rotation of the electron about its own axis, as discussed previously. Since the spin has been limited to just two values, represented by the two possible directions of rotation, the factor s must be limited to just two values. For electrons, s is taken as $+1/2$ or $-1/2$, representing spin up and spin down, respectively. These two values are indicated in (9.32).

The nuclear spin angular momentum is also in units of $(-e\hbar/2m)$, except that the mass of the nucleus is about 10^3 times that of the electron, so that the nuclear magneton is about $10^{-3}\beta$.

As an example of paramagnetic susceptibility, let us calculate the contribution of the susceptibility from the spin angular momentum of the electrons in a semiclassical approach. Consider a solid composed of a number of atoms with fixed magnetic moments μ_m. If there are n atoms per unit volume, it is expected that

$$M = n\langle\mu_m\rangle, \qquad (9.34)$$

where the angle brackets denote a suitable average over the orientations of the individual μ_m. This is a result of the classical theory of paramagnetism. In an external field, the additional energy due to the magnetic moment is

$$\Delta E = -\mu_m B \cos\theta = -\mu_m \mu_0 H \cos\theta = -\boldsymbol{\mu} \cdot \mathbf{B}, \qquad (9.35)$$

where θ is the angle between the directions of μ_m and H. In order to carry out the average, we need to realize that each state is weighted by $e^{-\Delta E/k_B T}$, the classical result of Fermi–Dirac statistics. We may see this as follows. The probability that a state at E is occupied is

$$P_{FD} = \frac{1}{1 + \exp\left(\dfrac{E - E_F}{k_B T}\right)} \sim \exp\left(-\frac{E - E_F}{k_B T}\right), \qquad (9.36)$$

while the probability that a state at $E + \Delta E$ is occupied is

$$P_{FD} = \frac{1}{1 + \exp\left(\dfrac{E + \Delta E - E_F}{k_B T}\right)} \sim \exp\left(-\frac{E + \Delta E - E_F}{k_B T}\right) \qquad (9.37)$$

for $E - E_F \gg k_B T$. Then the ratio of the occupation probabilities is

$$\exp\left(-\frac{\Delta E}{k_B T}\right). \qquad (9.38)$$

Then,

$$
\langle \cos\theta \rangle = \frac{\displaystyle\int e^{-\Delta E/k_B T} \cos\theta \, d\Omega}{\displaystyle\int e^{-\Delta E/k_B T} \, d\Omega}
$$

$$
= \frac{2\pi \displaystyle\int_0^\pi \cos\theta \sin\theta \exp\left(\dfrac{\mu_m \mu_0 H}{k_B T} \cos\theta\right) d\theta}{2\pi \displaystyle\int_0^\pi \sin\theta \exp\left(\dfrac{\mu_m \mu_0 H}{k_B T} \cos\theta\right) d\theta}. \qquad (9.39)
$$

Here, we have recognized that the solid angle differential is $2\pi \sin\theta\, d\theta$, since there is no azimuthal variation. To facilitate the computation, we make the substitution

$$y = \cos\theta, \qquad x = \frac{\mu_m \mu_0 H}{k_B T}, \tag{9.40}$$

and

$$\langle\cos\theta\rangle = \langle y\rangle = \frac{\displaystyle\int_{-1}^{1} e^{xy} y\, dy}{\displaystyle\int_{-1}^{1} e^{xy}\, dy} = \frac{\dfrac{d}{dx}\displaystyle\int_{-1}^{1} e^{xy}\, dy}{\displaystyle\int_{-1}^{1} e^{xy}\, dy} = \frac{d}{dx}\ln\int_{-1}^{1} e^{xy}\, dy,$$

$$\langle\cos y\rangle = \frac{d}{dx}\ln\left(\frac{e^x - e^{-x}}{x}\right) = \coth(x) - \frac{1}{x}. \tag{9.41}$$

Then, from (9.34), the magnetization is

$$M = n\mu_m\left(\coth x - \frac{1}{x}\right). \tag{9.42}$$

The term in parentheses is called the *Langevin function*. For $x \ll 1$ ($\mu_m H \ll k_B T$), this function may be approximated as $x/3$. In this approximation, the magnetization is

$$M = \frac{n\mu_0 \mu_m^2 H}{3k_B T}, \tag{9.43}$$

and the magnetic susceptibility is

$$\chi_m = \frac{n\mu_0 \mu_m^2}{3k_B T} = \frac{C}{T}, \tag{9.44}$$

where C is defined as the *Curie constant*

$$C = \frac{n\mu_0 \mu_m^2}{3k_B}. \tag{9.45}$$

If we carry out a complete quantum-mechanical treatment, we find a somewhat different result. Earlier, we pointed out that the magnetic moment due to spin was given by

$$\mu_s = g_s s \beta, \tag{9.46}$$

where β is the Bohr magneton $(-e\hbar/2m)$. For a single electron, with no orbital angular momentum, it is found that the energy change is due primarily to the spin angular momentum. Then μ_m is just μ_s and

$$\Delta E = \mp \mu_0 s H, \tag{9.47}$$

where the $\mp$ sign refers to spin up/down. Earlier, we observed that the energy of a quantum state was determined by the three quantum numbers k_x, k_y, and k_z. From the Pauli principle, each state so determined has states for the two possible electron spins. When a magnetic field is applied, the energy of one spin orientation is increased by the amount $\mu_s B$, while the energy of the other is reduced by this same amount. The spin angular momentum is now oriented either parallel or antiparallel to the magnetic field. Because of the difference in energy of these two possible orientations, it is expected that more of the electrons will be oriented with the lower energy. Therefore the total magnetization will be due to the small difference between the numbers of each orientation of the spin angular momentum. Suppose that we approximate the Fermi–Dirac statistics as

$$P_{\text{FD}} \sim e^{E_F/k_B T} e^{-E/k_B T}. \tag{9.48}$$

Then, for n_1 electrons in the upper state and n_2 in the lower $(n_1 + n_2 = n)$,

$$\frac{n_1}{n} = \frac{n_1}{n_1 + n_2} = \frac{e^{-(E_0 + \mu_s \mu_0 H)/K_B T}}{e^{-(E_0 + \mu_s \mu_0 H)/k_B T} + e^{-(E_0 - \mu_s \mu_0 H)/k_B T}} = \frac{e^{-x}}{e^x + e^{-x}}, \tag{9.49}$$

where $x = \mu_0 \mu_s H / k_B T$. Likewise,

$$\frac{n_2}{n} = \frac{e^{+x}}{e^x + e^{-x}}, \tag{9.50}$$

The projection of the upper state along the field direction is $-\mu_s$, while that of the lower states is μ_s. Thus the n_1 electrons in the upper state oppose the magnetic field, while the n_2 in the lower state aid the field with their magnetic moments. Then the total magnetization M is

$$M = (n_2 - n_1)\mu_s = n\mu_s \tanh(x). \qquad (9.51)$$

The function $\tanh(x)$ is different from the Langevin function found in the classical case. Even for $x \ll 1$, the results are different. In this latter case, it is found that $\tanh(x) \sim x$, and (9.51) differs from (9.43) by a factor of 3, as well as the orbital magnetic moment μ_m being replaced by the spin angular moment μ_s. This difference (the factor of 3) is not unexpected, since electron spin itself is not a classical result, and this factor arose from a classical angle averaging procedure.

9.4. FERROMAGNETISM

A substance is called *ferromagnetic* if it has a *spontaneous* magnetic moment M, that is, a magnetic moment exists even in the absence of an external magnetic field, as shown in Fig. 9.3. Ferromagnetic solids such as iron, cobalt, and nickel have very high values of the susceptibility, which are dependent on both magnetic field and temperature. This tells us that a ferromagnetic material has very large susceptibility χ_m. Above a certain temperature T_c, called the *Curie temperature*, the susceptibility is similar to that of paramagnetic materials, that is, it is relatively small. In general, the susceptibility for the ferromagnetic material can be written as

$$\chi_m = \frac{C}{T - T_c}, \qquad T > T_c, \qquad (9.52)$$

where C is the Curie constant from (9.45), and T_c is the Curie temperature. Below the Curie temperature, both the magnetization M and the magnetic flux density depend on the magnetic field in a nonlinear manner, and exhibit the phenomenon known as *hysteresis*

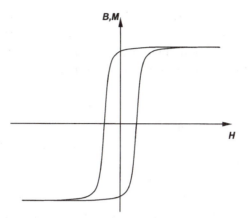

Figure 9.3. The variation of the magnetization and the magnetic flux density as a function of the applied magnetic field for a typical ferromagnetic material. The clear hysteresis indicates that a permanent magnetization of the material may be induced by applying a magnetic field.

(Fig. 9.3). The curve is different for increasing and decreasing H. A saturation of B and M occurs at large values of H. The saturation value M_{sat} of the magnetization, at high magnetic field intensities, corresponding to B_{sat}, depends only on the nature of the solid, that is, on the type of material and the purity of it. The maximum magnetization $M_{sat}(T)$, as a function of temperature, decreases to zero at $T = T_c$, so that ferromagnetism essentially disappears for temperatures above the Curie temperature. Some typical values for the Curie temperature are 1043, 1400, and 631 K, for Fe, Co, and Ni, respectively.

We may assume that there is an *internal* magnetic field intensity H_I, which is due to the presence of a magnetization M, and that these two factors are related by

$$H_I = \lambda M, \tag{9.53}$$

where λ *is the Weiss constant*, named after the founder of the present theory. Weiss supposed that the internal field H_I was due to the cooperative interaction of neighboring magnetic dipoles. To produce the observed results, all the dipoles in a solid must cooperate and be aligned in a parallel direction. Based on his assumptions, we may formulate a phenomenological theory that describes many of the experimental details. The approach here is very simplified, but it will serve to illustrate the phenomenon. From (9.51), when all the moments are aligned, the magnetization is

$$M = n\mu_m \tanh(x), \tag{9.54}$$

where μ_s has been replaced by the orbital magnetic moment μ_m, and $x = \mu_0\mu_m H/k_B T$ in this case. Since the internal field must also be included, H is replaced by $H + \lambda M$, and

$$x = \frac{\mu_m}{k_B T} B_{\text{Total}} = \frac{\mu_m}{k_B T}\mu_0(H + \lambda M). \tag{9.55}$$

Since the magnetization may exist without external fields, we wish to find the spontaneous magnetization. For $H = 0$, (9.55) becomes

$$x_0 = \frac{\mu_m}{k_B T}\mu_0\lambda M(T), \tag{9.56}$$

where an explicit temperature variation of the magnetization M is included. As T approaches 0, (9.54) gives

$$M(0) = n\mu_m. \tag{9.57}$$

Combining (9.56) and (9.57) yields

$$\frac{M(T)}{M(0)} = \frac{k_B T}{n\mu_m^2\mu_0\lambda} x_0. \tag{9.58}$$

But, from (9.54) and (9.57), we also find that

$$\frac{M(T)}{M(0)} = \tanh(x_0), \tag{9.59}$$

where x_0 is used for $H = 0$. The easiest way to solve the two equations is graphically, by plotting both (9.58) and (9.59), and finding their intercept, which is the point which satisfies both

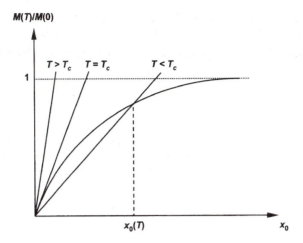

Figure 9.4. This figure shows that a net magnetization may only be induced in a ferromagnetic material at temperatures below the Curie temperature, T_c. The value of this magnetization is determined by the requirement that at a given temperature the magnetization should correspond to a solution of both Eqs. (9.58) and (9.59).

equations, and thus is the solution. This is shown in Fig. 9.4 for several different temperatures. As T approaches 0, $x_0 \to \infty$, and $\tanh(x) \to 1$, which is its maximum value. Thus $M(T)/M(0) \leqslant 1$. Below a critical temperature T_c, where (9.58) is tangential to (9.59) at $x_0 = 0$, two solutions exist. The nonzero solution corresponds to a spontaneous magnetization.

Thus, below a critical temperature in a ferromagnetic material, there can exist a spontaneous magnetization in the absence of an external magnetic field. Above T_c, the only solution is for $M(T) = 0$, and there is no spontaneous magnetization. For x near 0, (9.59) becomes

$$\frac{M(T_c)}{M(0)} \sim x_0, \tag{9.60}$$

and from (9.58) we have

$$x_0 = \frac{k_B T}{n\mu_m^2 \mu_0 \lambda} x_0, \tag{9.61}$$

so that the Curie temperature is given by

$$T_c = \frac{n\mu_m^2\mu_0\lambda}{k_B}. \tag{9.62}$$

To explain the fact that ferromagnetic materials are normally found in a demagnetized state, Weiss proposed a second postulate. The solid is assumed to be composed of a number of small magnetized domains, which are randomly oriented in the solid. Thus the net magnetization of the material may vary from zero to M_{sat}. Thus, the field can be used to orient the domains as a whole. Once the coercive force has been achieved, all the domains will begin to line up, yielding the saturation magnetization. To reverse this, the field must be turned around with a value greater than the coercive force in this opposite direction. The hysteresis of the $B-H$ curve results from the losses encountered when the magnetization domains are forcibly oriented by the external field. The losses are due to the inertia of the domains, and represent the energy that must be expended in orienting the domains into a coherent whole. In Fig. 9.5, we indicate a sketch of the domains, their random orientation and their oriented state. Generally, these domains relate to the grains of the polycrystalline magnetic material. In Fig. 9.6, a high resolution micrograph of the grain structure in a $Co_{0.35}Ag_{0.65}$ alloy is shown. For permanent magnets, we want to have a material that has a very high coercive force, so that it is difficult to change the internal magnetization.

For memory applications, however, a high coercive force means that significant energy is dissipated during the switching of the memory cell (the state is determined by the direction of the magnetization). In this application, the coercive force and saturation magnetization both should be relatively small, to keep the dissipation down, but still must be sufficiently high that the magnetized state is stable for a very long period of time, and can be sensed without destroying the state. In Fig. 9.7, a scanning magnetic microscope (variant of the atomic force and scanning tunneling microscopes) image of the bits in an IBM memory that has a bit density of 20 billion bits/in.[2] Material such as that used by IBM coats the rotating disks in the "hard drive" memory of a computer. These memories are cheaper, in terms of cost per bit, than semiconductor memories and, although slower in access time than semiconductor memories, are used for such archival storage applications.

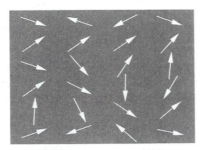

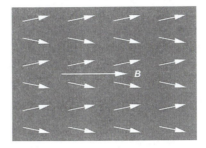

Figure 9.5. Domains of magnetization in a ferromagnetic material. The top figure shows randomly oriented domains in the presence of which the net magnetization of the material is zero. The bottom figure shows the situation where the domains have been aligned to yield a net magnetization.

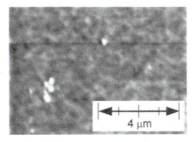

Figure 9.6. Observation of antiparallel magnetic order in weakly coupled Co/Cu multilayers. This image was taken using a technique known as scanning electron microscopy with polarization analysis (SEMPA), in which the magnetization of a material is imaged by measuring the spin polarization of secondary electrons emitted in a scanning electron microscope. *J. A. Borchers, J. A. Dura, J. Unguris, D. A. Tulchinsky, M. H. Kelley, C. F. Majkrzak, S. Y. Hsu, R. Loloee, W. P. J. Pratt, and J. Bass, Physical Review Letters* **82**, *2796–2799 (1998). Reproduced with permission.*

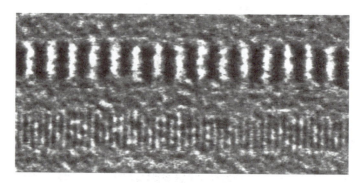

Figure 9.7. Scanning magnetic microscope image of bits in an IBM memory that has a bit density of 20 billion bits/in.^{2} (lower line). For comparison, a test pattern written at 6.4 billion bits/in.^{2}, slightly greater than typical disk-drive densities, is also shown. *Picture reproduced courtesy of IBM. Copyright IBM corporation.*

9.5. GIANT MAGNETORESISTANCE

Just as in semiconductors structures, the magnetic properties of a thin magnetic material (or nonmagnetic material) can be modified by creating heterostructures. Consider the application shown in Fig. 9.8. The two Fe layers are ferromagnetic, in that the Curie temperature of Fe is 1043 K, as mentioned above. In the construction of Fig. 9.8, the magnetization of the two iron layers is designed to lie in opposite directions, while the Cr layer is unmagnetized. If the latter is less than, for example, 1 nm thick, then the magnetization of the two layers can couple (in effect, it tunnels through the nonmagnetic Cr layer) and create an antiferromagnetic ordering. In this regard, the presence of the Cr layer significantly raises the resistance of the overall structure for current flow parallel to the heterostructure plane. However, when a magnetic field is applied, in the plane of the heterostructure, it reverses the magnetization direction in one of the two films of Fe. This gives a cooperative interaction which significantly reduces the resistance of the Cr layer. This produces a *negative* magnetoresistance, since the resistance is decreased with the applied magnetic field. Normally, metals show almost no magnetoresistance, but these heterostructures can have resistance changes that can range from a few percent to almost a factor of 2, when 30–60 multiple layers are used. For this reason, the effect is known as *giant*

Cr   Fe   Cr

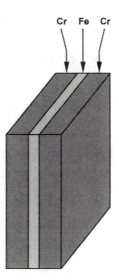

Figure 9.8. A magnetic "heterostructure" that may be used to study GMR (giant magneto-resistance) effects.

magnetoresistance. In Fig. 9.9(a), we show the magnetoresistance for a simple structure (of Fig. 9.8), where the change is about 1.5%. On the other hand, the resistance itself is plotted in Fig. 9.9(b) for several multilayers of Fe–Cr films, where a much larger effect is seen. This GMR effect is finding new applications as the read head for magnetic disks. Normally, the Hall effect has been used, but these new GMR heads provide more sensititivity and a larger signal from the passage of the magnetic bit, and this means a better signal-to-noise ratio for the system. They also appear to be more stable with time, especially for digital applications.

9.6. MAGNETIC MEMORY

From the earliest computers, magnetic materials have been used for storage of the binary bits. The magnetization is a perfect medium for this storage as the two directions of polarization correspond nicely to the "0" and "1" of binary arithmetic. The first such use occurred

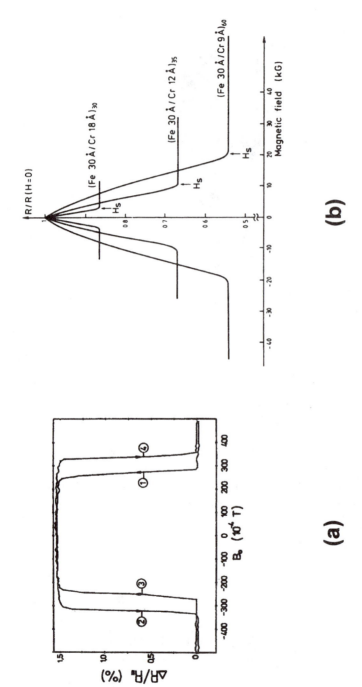

Figure 9.9. GMR effect in different structures (a) Weak GMR effect for the simple structure of Fig. 9.8. Note that the maximum magneto-resistance effect here is about 1.5%. G. Binasch et al., *Phys. Rev. B* **39**, 4828 *(1989)*. *reproduced with the permission of the authors.* (b) Much larger GMR effects may be observed in multilayer films, in this case realized from iron and chromium. M. N. Baibich et al., *Phys. Rev. Lett.* **61**, 2472 *(1988)*. *Figure reproduced with the permission of the authors.*

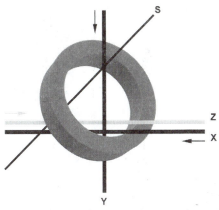

Figure 9.10. A schematic view of a magnetic core, once used as the main memory in computers. The various wires are used to write a magnetization (current running both through Z and X) and to read the state (current through either Z or X and sensing with S).

both as rotating magnetic drums (in which the magnetic material was on the surface of the rotating drum) and magnetic core memories, which were used for random access memories. Such a single bit core memory is shown in Fig. 9.10. The assortment of wires are needed to write the memory bit and then to read it in a nondestructive manner. Of course, such random access memories disappeared with the advent of MOS capacitor–transistor memory cells, such as those described in Chapter 6.

Today, however, the so-called "mass" memory still remains magnetic. Instead of large rotating drums, we now use rotating "disks," which are read by movable heads. The speed of this memory is increased by increasing the rotational speed of the disc and the lateral motion of the heads. A multipledisk memory structure is sketched in Fig. 9.11. The reason that this memory persists is cost per bit. Magnetic bits can be as small, or perhaps smaller than a transistor bit. But, the cost per bit remains about a factor of ten less than that for MOS memory. As a result, the mass memory is quite likely to remain magnetic in the forseeable future.

A new magnetic memory, which has promise to replace MOS memory, is based on the giant magnetoresistance structures of the last section. Here, one creates a magnetic tunneling structure within the GMR layers, and the current flow is configured to be normal to the layers, as shown in Fig. 9.12. Hence, in the high-resistance state,

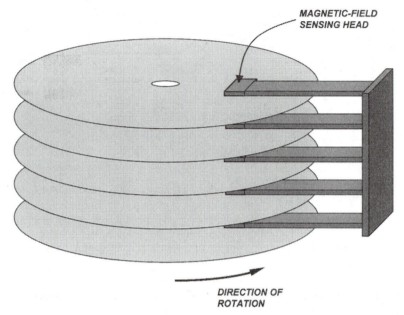

Figure 9.11. A five "platter" magnetic disk memory. The heads are moved with the cantilevers to read a particular region on the disk as it rotates.

we have one binary state, but when we switch to the low-resistance state, we have the opposite binary state. The state is sensed by the current through the structure and the transistor. These structures have been named magnetic random access memory, or MRAM. However, only time will tell whether or not these devices actually become a key component in computers.

PROBLEMS

1. Show that the magnetic flux density B at a distance r from an infinitely long straight wire carrying a current I is given by

$$B = \mu_r \mu_0 \frac{I}{2\pi r}.$$

What is the direction of the flux lines?

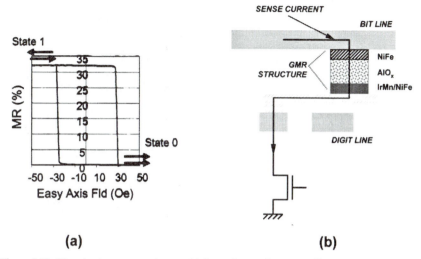

Figure 9.12. The giant magnetoresistance (a) is used to make a tunneling magnetic memory. In one state, current can tunnel through the GMR structure to the transistor (b). In the other state, the current is kept small. *Picture courtesy of H. Goronkin, Motorola Laboratories.*

2. Consider a spinning spherical shell of total charge e and total mass m. Both the charge and the mass are uniformly distributed over the surface of the shell. Show that the ratio of the magnetic moment to the angular momentum is $e/2m$.

3. The Curie temperature of iron is 1043 K. It may be assumed that iron atoms have a magnetic moment of two Bohr magnetons per atom. Calculate the Weiss constant predicted in (9.61). What are the Curie constant and the saturation magnetization at 0 K?

4. Consider a cylindrical shell with a uniform azimuthal surface current density of 0.1 A/cm^2. The shell is 10^{-4} in. thick, has a radius of 2 cm, and is 0.5 cm long. What is the total magnetic dipole moment?

CHAPTER 10

Superconductivity

In 1911, Kammerlingh Onnes was busy investigating the low-temperature properties of materials in his laboratory in Leiden, the Netherlands. Only 3 years earlier, he had been the first person to be successful in the liquefaction of helium at 4.2 K, and was anxious to test various materials at this low temperature. At this time, he was investigating the resistance of mercury. As he cooled the mercury below 4.2 K, a strange phenomenon occurred. Although, by an extrapolation of the 0°C (273 K) resistance of 173 Ω, he expected the resistance to drop to less than 40 Ω, at 4.2 K the resistance of the mercury abruptly dropped to an exceedingly low value. In fact, at 3 K the resistance was measured to be less than $10^{-6}\,\Omega$. The phenomenon observed by Onnes is known today as *superconductivity*, and has been observed in over one-fourth of the elements of the periodic table, as well as in numerous compounds and alloys. The onset of superconductivity is considered to be a thermodynamic phase transition of the electronic material. The temperature at which the transition occurs is called the *transition temperature*, and below this temperature the material is superconducting.

In the ensuing years, not much was done, primarily due to the scarcity of liquid helium. It was not until after the second world war, when liquid helium could be produced relatively cheaply and in large quantities, that interest in superconductivity and its extremely useful properties was really aroused.

In the first part of this chapter, we shall discuss the experimentally observed properties of superconductors. Although the details are clearly beyond the scope of this text, we shall introduce the important aspects of the theoretical basis of superconductivity. A later section will deal with electronic devices utilizing superconductivity. These are primarily based on the tunnel junction, and find usage both as possible elements in electronic computers and as active microwave devices. The Josephson effect in tunnel junctions has put forth the possibility of millimeter-wave generators and detectors utilizing this effect, although the superconducting quantum-interference device (SQUID) is perhaps the most widely used application as it finds use as a sensitive magnetometer. At the end of the chapter, we will talk about the newer, so-called, high-T_c materials which hold promise for superconducting behavior closer to room temperature.

10.1. PROPERTIES OF SUPERCONDUCTORS

The most easily observed characteristic of superconductors is the *transition temperature* of the material. This temperature, T_c (not to be confused with the Curie temperature of Chapter 9), is the temperature below which superconductivity occurs in the material. One can readily observe the transition temperature when one plots the resistance versus the temperature, as in Fig. 10.1. The temperature scale in the figure is normalized to the transition temperature, T_c. There is a slight rounding of the curve just above T_c, and the enhanced conduction in this region is often called *paraconductivity*, in analogy with paramagnetism, or the enhancement of magnetism, which was discussed in Chapter 9. This enhanced conductivity is attributed to thermal fluctuations at the transition temperature, and we can think of them as very small regions, which are beginning to exhibit superconducting behavior. Superconductivity of the entire sample does not occur, since the small regions are superconducting for only short periods of time and are generally unconnected. In a sense, these are like the domains exhibited in magnetic field. Only a few of the domains are superconducting, and then only for short periods of time.

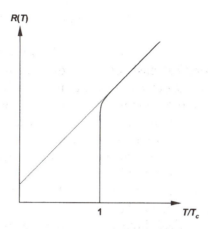

Figure 10.1. Schematic illustration of the superconducting transition that can occur in certain metals at a critical temperature T_c.

The drop in resistance at $T = T_c$ is exceedingly sharp, and occurs over a small fraction of a degree, so that the onset of superconducting behavior is easily observed. The transition temperature is a fundamental property of superconductors. Some typical transition temperatures for elements and compounds are shown in Table 10.1.

As mentioned, only about one-fourth of the elements of the periodic table exhibit superconductivity. Such elements as silver and copper are not superconductors at the lowest temperatures measured. One might ponder why this is true. In fact, it seems that the higher

TABLE 10.1. Superconducting Transition Temperatures

Element	T_c (K)	Compound	T_c (K)
Pb	7.2	Nb_3Sn	18.05
V	3.72	V_3Ga	16.5
Ti	0.39	V_3Si	17.01
Nb	9.1	NbN	16.0
Ta	4.48	InSb	1.9
Hg	4.15	Nb_3Al	17.5
Zn	0.85		
Sn	3.72		

the conductivity of a material in the normal state, the poorer this material is as a superconductor. This implies that if a material is a good conductor in the normal state ($T > T_c$), it probably has a very low transition temperature or does not exhibit superconductivity at all. As we shall see in the latter part of this chapter, we can understand this effect by thinking about the dominant interaction mechanism leading to superconductivity. This interaction, at least in the pure compounds, is generally conceded to be the electron–lattice interaction. The stronger this interaction is, the more likely a material is to be a superconductor. However, this is just the dominant scattering mechanism in normal metals. If the electrons are strongly scattered and the conductivity is low, this interaction is strong, and the conductivity is low. Hence the material is a poor conductor.

Although the evidence for the electron–lattice interaction being the chief mechanism leading to superconductivity is strong, there is evidence that at least one other interaction leads to it. This evidence is derived from measurements of the isotope effect. In the electron–lattice interaction, the strength of the interaction (and the vibration of the atoms in the lattice) is a function of the mass of the lattice atoms. If this is the proper interaction, then the critical temperature, T_c, should vary inversely as the mass of the lattice atoms, or

$$M_{atom} T_c = \text{constant}, \tag{10.1}$$

By comparing superconductors, whose lattices are prepared from isotopes of the same atoms, experiments have strongly confirmed this prediction. This was an early success of theoretical predictions of superconductivity. However, at least one element has been found that apparently does not obey this isotope effect. This is the element rhenium ($T_c = 1.7$ K). If rhenium indeed violates the rule, then this must constitute evidence that there is at least one other interaction mechanism that leads to superconductivity. It is not clear what this other mechanism is, and this question has remained unanswered. For the most part, though, the isotope effect is strong support for the electron–lattice interaction as the primary interaction leading to superconductivity.

A magnetic field can be used to destroy superconductivity. If a magnetic field is applied to a superconductor, then, for $H > H_c$,

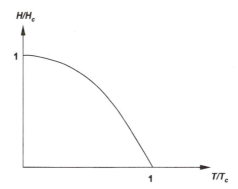

Figure 10.2. Variation of the critical magnetic field required to destroy superconductivity as a function of temperature.

a critical value of the magnetic field which is different for different superconductors, the superconductivity is destroyed and the material reverts to the normal state. It is generally found that H_c is a function of temperature as

$$H_c = H_{c0} \left[1 - \left(\frac{T}{T_c} \right)^2 \right], \qquad T < T_c, \qquad (10.2)$$

where H_{c0} is the *critical field at absolute zero*. The critical field at $T = 0$ K is a function of the critical temperature. The additional magnetic energy of the electrons serves to break the superconducting bond interaction, just as thermal energy does. The effect of both types of additional energy terms leads to a result like (10.2).

The critical magnetic field will also limit the amount of current which a superconductor may carry. Since the current gives rise to a magnetic field, the current carried by a superconductor must be less than that which would produce H_c. For a higher current than this, the magnetic field produced by the current destroys superconductivity and the material reverts to the normal state. Figure 10.2 shows a plot of the critical magnetic field as a function of temperature (10.2). Such a curve is followed very closely by most superconductors.

10.2. THE MEISSNER EFFECT

The onset of superconductivity leads to a phenomenon called the *Meissner effect*. If a superconductor is cooled below T_c, the magnetic flux lines are expelled from the material. Thus, within a superconductor, the magnetic flux density is zero, $B = 0$. In Chapter 9, we called a material in which the relative permeability $\mu_r < 1$ *diamagnetic*. Superconductors exhibit perfect diamagnetism, $\mu_r = 0$. The magnetization which results from the applied field must completely oppose this applied field. From (9.14), we may deduce that

$$B = \mu_0(H + M) = 0. \tag{10.3}$$

This leads to

$$M = -H, \qquad \chi_m = -1, \tag{10.4}$$

And

$$\mu_r = 1 + \chi_m = 0. \tag{10.5}$$

In Fig. 10.3(a), the magnetization M is plotted as a function of the field intensity H. For magnetic fields which exceed H_c, the material reverts to the normal state, and the magnetization of the material reverts to the type displayed in the normal state. Perfect diamagnetism, for fields below H_c, is displayed in pure bulk specimens of many materials. Materials that display this perfect diamagnetism are called *type I*, or *soft* superconductors. The values of H_c for such materials are very low, and the materials are not useful for magnetic-effect devices.

Other materials display a magnetization like that of Fig. 10.3(b). These materials, which are called *type II superconductors*. These generally tend to be transition metals or alloys. Here, the perfect diamagnetism begins to decay at a given level of external field intensity. Perfect diamagnetism occurs only up to H_{c1}, the lower critical field, and then the magnetism falls off to zero at a value of H_{c2}, the upper critical field. The value of H_{c2} can be very high, and these materials are useful for making superconducting magnets. A

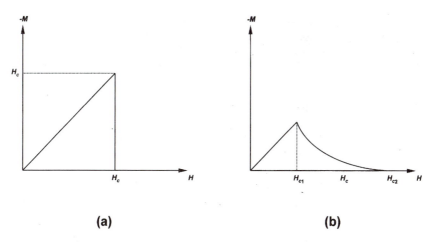

Figure 10.3. Variation of magnetization as a function of the applied magnetic field for (a) type I and (b) type II superconductors. Note the existence of *two* critical magnetic-field scales for the type II superconductor.

hard superconductor is a type II with magnetic hysteresis, an effect generally produced by mechanical treatment of the material.

A number of the preceding experimental observations are consistent with the existence of an *energy gap* associated with superconductivity. Theoretically, the electrons that contribute to superconductivity do so as paired electrons. This set of two electrons is known as a *Cooper pair*, and is formed from two electrons that have opposite spin angular momentum. Thus, if one of the electrons is spin up, the other is spin down. Normally, electrons repel one another because of Coulomb forces. But at low temperatures, an additional interaction between the electrons, in which they interact with the lattice, leads to a weak attractive force between the electrons. This leads to a pairing of the electrons. As will be seen later, this pairing of the electrons results in a lower energy state than would result from the electrons remaining unpaired. This, in turn, results in an energy gap of the electrons. The observed energy gap, though, would correspond to $E_G = 2\Delta$, since each of the paired electrons must receive an additional energy of Δ to break the pairing bond.

At absolute zero, all the conduction electrons are paired. At high temperature, some of the pairings are broken by thermal agitation, so that some normal electrons are excited across the energy gap in much the same way that electrons are excited across the band gap in semiconductors. For temperatures slightly above T_c, all the pairs are broken and the material exhibits normal resistivity. In the superconducting state, a mixture of normal electrons and superconducting paired electrons is present in the material. The normal electrons may be observed in tunneling experiments, such as we discuss in a later section of this chapter.

The energy gap also varies with temperature, decreasing with increasing T until, at T_c, the energy gap is zero. The general behavior of the gap near T_c is given by

$$E_G = E_{G0}\left(1 - \frac{T}{T_c}\right)^{1/2}, \tag{10.6}$$

where E_{G0} is the value of the gap at zero temperature. As a first approximation, the energy gap at zero temperature is related to the energy gap at the transition temperature as $E_{G0} = 3.5k_B T_c$. This result is not exact, but is approximately correct for a wide range of superconductors.

The existence of the energy gap can be readily demonstrated by measurements at microwave or infrared wavelengths. At absolute zero, photons with energy $\hbar\omega < E_G$ are not absorbed, while those with $\hbar\omega > E_G$ cause transitions across the energy gap and result in absorption of photons. The photon energy is absorbed by the electron pair, and this additional energy is sufficient to break the bond of the pair. Figure 10.4 shows such an absorption curve for a typical superconductor. For temperatures above zero, some absorption occurs due to the presence of a few normal electrons. Since these normal electrons are already excited across the gap, they are free to absorb any photon that comes along.

A somewhat different behavior is observed if the superconductor is fabricated as a thin film. If the films are very thin, the material does not exhibit perfect diamagnetism. As we shall see in the next section, the film must be of a certain thickness before it exhibits the properties of bulk superconductors, such as perfect diamagnetism. In addition, in a thin film, the magnetic field has very little effect and the critical field is much higher than for the bulk material.

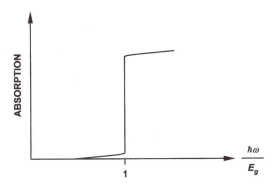

Figure 10.4. Variation of the microwave absorption as a function of frequency for a typical superconductor. As the photon energy becomes comparable to $E_g = 2\Delta$ a dramatic increase in absorption occurs as the microwave radiation begins to break up the Cooper pairs.

An exceedingly useful property in superconductors is the presence of *persistent currents*. Consider, for example, a superconducting ring. If a current is induced in the material in the normal state, and the material is then cooled below T_c, the current in the ring will persist for a very long time. Since the material has no resistance in the superconducting state, there is no decay mechanism for the current. Such a mechanism is exceedingly useful in superconducting magnets. If a fixed value of magnetic field is required, the superconducting solenoid is energized to the proper current value by an external power source. Then the current leads are shorted together at the magnet by a second superconductor. This forms a superconducting loop, and the external supply can be turned off, while the circulating current in the loop remains unchanged.

10.3. THE LONDON EQUATIONS

As we pointed out earlier, the Meissner effect is not fully observed in thin-film superconductors. This can be understood on the basis of one of the earliest theoretical descriptions of superconductivity. In Chapter 5, we found that the force equation for electrons could be given as

$$m\frac{d\mathbf{v}}{dt} = -e\mathbf{F}. \tag{10.7}$$

In the case of a normal metal, we can simply integrate (10.7) up to some time τ, where τ is the *mean collision time*. In a resistanceless state, however, this cannot be done, since such a collision time cannot be defined. Since $J = -nev$, we can rewrite (10.7) as

$$\frac{d\mathbf{J}}{dt} = -ne\frac{d\mathbf{v}}{dt} = \frac{ne^2}{m}\mathbf{F}. \tag{10.8}$$

Because of the accelerating nature of the electric field, it is clear that (10.8) can be satisfied only when $\mathbf{F} = 0$, or when there is zero resistance, even though a current may flow. This is, of course, the situation of superconductivity. Moreover, because of the magnetic field produced by the current, the current must be related to this field by the magnetization. For this effect, Hans and Fritz London postulated that the current should be related to the magnetic field as

$$\nabla \times \left(\frac{m\mathbf{J}}{ne^2}\right) = -\mu_0\mathbf{H}. \tag{10.9}$$

For a superconductor filling the plane, $x > 0$, a tangential magnetic field in the y direction, and a surface current in the superconductor directed in the z direction,

$$\frac{m}{ne^2}\frac{\partial J_z}{\partial x} = \mu_0 H_y, \tag{10.10}$$

which is a specialization of Faraday's law using (10.7). Equation (10.10) is, in fact, the *London hypothesis for superconductivity*.

One can define a vector magnetic potential which is related to the magnetic field as

$$\mathbf{B} = \nabla \times \mathbf{A}, \tag{10.11}$$

and we find that

$$\mathbf{J} = -\frac{ne^2}{m}\mathbf{A}. \tag{10.12}$$

This last equation is often called *the* London equation. The pair of equations (10.12) and (10.8) are often grouped together and called the London equations. Nevertheless, the magnetic field intensity and current must still satisfy *Ampere's law,*

$$\mathbf{J} = \nabla \times \mathbf{H}. \tag{10.13}$$

When we introduce (10.13) into (10.9), we can write the resultant relationship for the magnetic field intensity in terms of a wave equation as

$$\nabla^2 \mathbf{H} = \frac{ne^2\mu_0}{m} \mathbf{H}. \tag{10.14}$$

Consider the case of a magnetic field applied to the superconductor in the plane $x > 0$. Then, with the magnetic field parallel to the surface of the superconductor, we seek solutions to the equation

$$\frac{d^2 H_y}{dx^2} = \frac{1}{\lambda_L^2} H_y, \tag{10.15}$$

where

$$\lambda_L = \sqrt{\frac{m}{ne^2\mu_0}} \tag{10.16}$$

is the London penetration depth. The quantity λ_L gives the average depth of penetration of the magnetic field into a superconductor, and is the foremost result of the London formulation. If the thickness of a superconductor is large compared to λ_L, the magnetic field is negligible in the interior and the Meissner effect occurs. If the thickness is small compared to the penetration depth, the magnetic field easily penetrates the film and there is very little magnetization.

Thus, the magnetic field has little effect on thin superconducting films if $d \ll \lambda_L$, where d is the thickness of the superconductor. The success of the London theory lies in its prediction of the effects of magnetic field and thickness in thin films. It does not, however, contain an adequate treatment which leads to justifications of the

energy gap and its relation to the transition temperature. The
London theory is thus primarily a phenomenological theory, with its
main validity with respect to the Meissner effect.

10.4. THE BCS THEORY

A general theory of superconductivity—the BCS theory—was developed in 1957 by Bardeen, Cooper, and Schrieffer. The basic
developments of their theory show that an attractive force between
the electrons can lead to a lower energy state at low temperatures,
and, hence, can lead to an energy gap. For the attractive force, their
theory suggests a second-order interaction between electrons and the
lattice. One electron interacts with the lattice; this interaction de-
forms the lattice slightly in the neighborhood of the electron. This
deformation is much like the modification of the density of the
electrons near an ionized impurity that leads to electron screening
effects over a length of the order of the Debye length. In the BCS
theory, a similar effect occurs between the electron and the deformed
lattice. A second electron encounters the deformed lattice and interacts with it. Thus, in effect, the second electron interacts with the first,
but does so through the lattice interaction. Effectively, one could say
that the deformed lattice shields the electron's repulsive force so that
the second electron is attracted to the deformed area, and hence to
the first electron.

The idea of the London penetration depth evolves from the BCS
theory as a natural consequence. From the BCS theory, the critical
temperature is found to be a function of the electron density-of-states
function $\rho(E)$ at the Fermi energy level, and of the *interaction
potential between the electrons and the lattice*. We can find the latter
quantity, U, from the resistivity of the material if the electron–lattice
scattering is the dominant scattering mechanism for the electrons. An
important aspect is that the two electrons must have opposite spin,
otherwise the Coulomb interaction between them is just too strong
for them to form a pair. The attractive force between these two
electrons, with opposite spin, *lowers* the energy of the pair. Thus, it
costs energy to break the pair and a gap opens between the paired
electrons and the un-paired electrons, as shown in Fig. 10.5. The pair

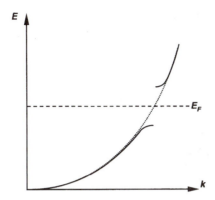

Figure 10.5. In the superconducting state, an energy gap opens near the Fermi surface and separates the superconducting and normal states.

of electrons is called a Cooper pair, and the gap is

$$E_G = 2\Delta, \tag{10.17}$$

where Δ is the energy lowering of each of the two electrons in the pair. The details of the BCS theory are beyond the scope of this book, but we can state the pertinent result for the critical temperature as

$$T_c = 1.14 T_D \exp\left(-\frac{1}{U\rho(E_F)}\right), \tag{10.18}$$

where T_D is the *Debye temperature* of the material, a characteristic constant for each material. From the form of (10.18), it is apparent that for a strong electron lattice interaction, U is large and T_c is also relatively large. But this is just the case for a small mean collision time τ in the conductivity ($\sigma = ne\mu = ne^2\tau/m^*$), which leads to a poor conductor. Thus, (10.18) is consistent with experimental observations.

Normally, when a gap opens, the electrons cannot carry any current as they cannot gain any energy. This was important in the band theory of semiconductors in Chapter 5. In fact, superconductors carry current without dissipation, so there is no need for the paired electrons to gain energy. In fact, the pair will hold together until a large amount of energy, given by the gap, is injected into the pair by some mechanism. Although the formal treatment is complex, a simple argument can show that for resistance to occur, there must be a critical minimum velocity, which corresponds to the presence of an energy gap. If the velocity of the electrons is less than the value of the critical velocity, no dissipation, or energy loss, occurs, and the material has zero resistance.

Consider the interaction of an electron with the lattice through a collision. If the incident electron has a velocity $-v$ and an energy $m^*v^2/2$ before the collision, it may be assumed that it loses in the collision an amount of energy given by Δ_k. With respect to the electron, the lattice is moving at a velocity v. It is convenient to require the lattice energy to be conserved, so that if the lattice moves at a velocity v' after the collision, conservation of energy requires that

$$\tfrac{1}{2}Mv^2 = \tfrac{1}{2}Mv'^2 + \Delta_k, \qquad (10.19)$$

where M is the average mass of the lattice atom. Conservation of momentum is also required, or

$$p = p' + \Delta p, \qquad (10.20)$$

using $\Delta p = \hbar k$ as the momentum change in the collision. We can combine these two equations as

$$\frac{1}{2}Mv^2 = \frac{1}{2}M\left(\frac{\mathbf{p} - \hbar\mathbf{k}}{M}\right)^2 + \Delta_k$$
$$= \frac{1}{2}Mv^2 + \frac{\hbar^2 k^2}{2M} - v\hbar k \cos\theta + \Delta_k, \qquad (10.21)$$

or

$$v\hbar k \cos\theta = \Delta_k + \frac{\hbar^2 k^2}{2M}. \qquad (10.22)$$

In these last two equations, θ is the angle between the direction of v and that of k. If the mass of the lattice is very large, the second term on the right is very small and can be neglected. The smallest v that satisfies the above equation occurs for $\cos \theta = 1$, so that the critical velocity is

$$v_c = \frac{\Delta_k}{\hbar k}. \tag{10.23}$$

Superconducting currents can flow without resistance, provided that the current is less than the critical value. The minimum value of Δ_k is the energy gap Δ of the material, so that quite large currents can flow. For the momentum, normally the value at the Fermi energy level is sufficiently accurate for metals, or

$$\frac{\hbar^2 k^2}{2m^*} = E_F. \tag{10.24}$$

In Pb, for example, the Fermi energy is 5.2 eV and the superconducting transition temperature is 7.2 K. Hence, the Fermi wave vector k is 1.17×10^8 cm^{-1}, and the energy gap is 2.17 meV. Thus, the critical velocity is 2.82×10^4 cm/s, which is relatively small when compared to semiconductor velocities. Nevertheless, with an electron density of 5.4×10^{22} cm^{-3}, this corresponds to a current of 2.4×10^8 A/cm^2. Metals and superconductors make up for their low velocities with the large number of electrons that contribute to the conduction. We note, however, that these values are at 0 K, and the gap and critical velocity are reduced as the temperature is raised toward the transition temperature according to (10.6).

10.5. SUPERCONDUCTING TUNNELING

If two superconductors are separated by a thin nonsuperconducting, or insulating, layer, such as an oxide, it is possible for electrons to tunnel through the insulator between the superconductors. For the present, let us assume that $T = 0$ K, so that all the electrons present are paired. In the absence of applied bias, no current flows and there is no tunneling, because all available states are filled on both sides of

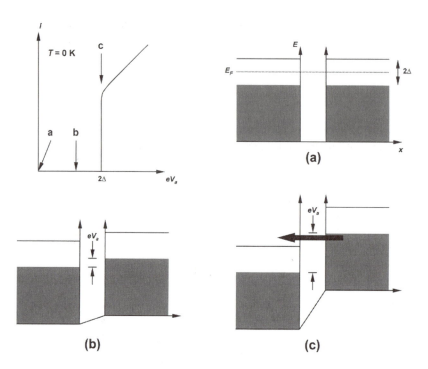

Figure 10.6. The current–voltage characteristic of a Josephson junction at 0 K and the associated energy band diagrams. (a) No applied voltage, (b) small applied voltage $eV_a < 2\Delta$, and (c) large applied voltage $eV_a > 2\Delta$.

the barrier. If voltage is applied between the two superconductors, the energy of one superconductor is raised relative to the other (see Fig. 10.6). Still no current can flow for $eV < 2\Delta = E_G$, due to the energy gap. There are no allowed energy states in the gap region. Only when $eV > 2\Delta$ will current flow, since paired electrons from one superconductor can tunnel into a region above the energy gap in the second superconductor. However, the applied voltage also breaks the pair bond, as well as causing tunneling. This is because the paired electrons tunnel into a region which lies above the superconducting gap. Hence, the bias energy is converted into the energy needed to break the pairing, with any residual going into kinetic energy of the unpaired electrons. In this sense, the tunneling transition is somewhat akin to the process of photoemission, where the superconducting gap plays the role of the work function.

If $T \neq 0\,\mathrm{K}$, a small number of unpaired electrons are available above the gap and are free to tunnel at any applied voltage. This constitutes a small tunneling current for $eV < 2\Delta$, as shown in Fig. 10.7. The rapid increase in current that occurs for $eV > 2\Delta$ still arises, as this is the onset of the pair breaking with the resulting single particle tunneling. In this high voltage regime, the current is essentially that of a metal-oxide–metal tunneling junction.

If the barrier, normally an oxide layer, is very thin, say 1–2 nm, then it is possible for the wave functions of the paired electrons to extend through the barrier. The Cooper pair is a very coherent entity, so that it's wave function extends for quite some distance. When the wave function can extend through the tunnel barrier, tunneling of superconducting electron pairs can occur at zero applied bias. That is, a current can flow with no applied bias. This is known already from the properties of superconductors, but the new effect is tunneling without any bias. This phenomenon is known as the *dc Josephson current*. There is a maximum current density, at which the pairs are broken by the induced magnetic field, at which point the voltage jumps to a value on the normal tunneling curve, as shown in Fig. 10.8. Because of the coherence of the wave function, an applied

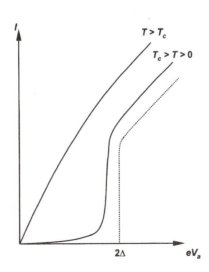

Figure 10.7. Schematic illustration of the current–voltage characteristics of a Josephson junction at several different temperatures. The dotted line shows the 0 K result.

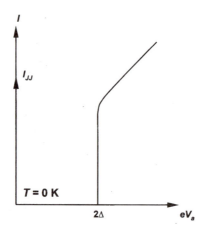

Figure 10.8. The current–voltage characteristic of a Josephson junction at 0 K. A Josephson tunneling current can flow at *zero* applied voltage and is denoted as I_{JJ} in this figure.

magnetic field can produce a significant modulation of the Josephson current. This leads to a variation as

$$I_J(B) = I_{J0} \cos\left(\frac{eBA}{h}\right), \qquad (10.25)$$

where $BA = \Phi$ is the flux flowing through the area A of the junction. The quantity h/e is the fundamental unit of flux, and in a quantum system, this flux can become quantized.

The equivalent system to the quantum well that we discussed in Chapter 3 is the superconducting quantum interference device (SQUID), in which the Josephson junction is embedded in a small current loop. Here, the flux contained within the loop is quantized in units of $\Phi_0 = h/e$. In Fig. 10.9, we illustrate a SQUID. If the area is 1 mm^2, then the magnetic field is quantized in units of 41.27 gauss (4.127×10^{-3} T). For larger areas, the quantized unit is correspondingly smaller. If the area of the loop is increased significantly, then the magnetic field for one oscillation is reduced accordingly. In general, there are two types of SQUID. In the case of the *dc* SQUID, there are usually *two* junctions in the loop. A dc current is passed through the loop, and the value of this current is set just above the

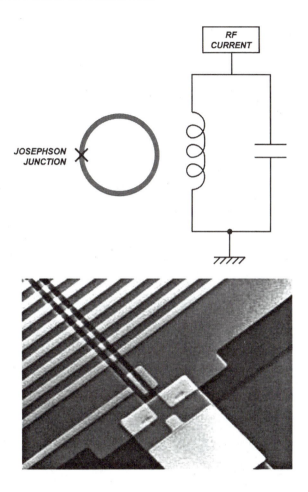

Figure 10.9. A *superconducting quantum interference device* circuit (or *SQUID*) contains a ring of superconducting metal that features an ultranarrow region, or *weak link*, on its circumference. The photograph shows an image of a niobium *SQUID*. Some turns of the input coil can be seen as well as the two Josephson junctions (crosslike structures in the lower half of the picture). *Image reproduced with the kind permission of Dr. Michael Mück, Institute of Applied Physics, Justus-Liebig-University Giessen, Germany.*

sum of the two Josephson currents of the two junctions. This means the devices are now operated in the resistive regime. As the Josephson currents are modulated by the magnetic field, the resistance change can be measured easily. In the ac SQUID, however, a single junction is placed in the loop. By coupling the SQUID to a sensing loop (typically a resonant *LC* circuit), background magnetic fields cause a change in the ac current in the loop, and this is coupled to the resonant circuit. The resulting voltage across the circuit can be measured, which means that the coupled magnetic field can be measured to a very small part of a tesla. These devices are used extensively for sensitive magnetometers.

There is a second Josephson effect, called the ac effect, which is also observed. It too relates to the coherent wave function within the tunneling junction. The phase difference in the wave function on the two sides of the barrier is a function of the potential difference across the barrier. In this regard, the phase difference can be considered to be equivalent to a magnetic flux, so that the analog of Faraday's law can be imposed. That is, the phase difference relates to the potential through

$$V = \frac{h}{e}\frac{d\varphi}{dt}. \tag{10.26}$$

Here, φ is the phase difference between the coherent states on either side of the tunnel junction. The phase will therefore oscillate when a voltage is applied to the junction. This is equivalent to a small oscillating magnetic dipole, so that radiation arises from the tunnel junction. Hence, the frequency is related to the voltage through

$$eV = \hbar\omega. \tag{10.27}$$

In a Pb–oxide–Pb junction, the energy gap is 2.17 meV, so that we can apply easily a 1 mV bias to the junction. This leads to a radiation frequency of 2.42×10^{11} Hz, or 0.24 THz. This radiation has a wavelength of 1.24 mm, or well beyond the microwave regime. However, the frequency is so intrinsically coupled to the voltage applied, that this device has become the frequency standard for most countries in the world.

At the same time, the Josephson device is a good detector of radiation, with the resulting voltage given by the frequency through (10.27). A secondary result is that the single-particle tunneling, which provides the current shown in Fig. 10.7 when $eV < 2\Delta$, is a nonlinear process. As a result, the single-particle tunneling current will show a detected current proportional to the radiation of the Josephson junction. This current appears as steps in the current, with each step corresponding to an increase in the number of photons that are absorbed during the single-particle tunneling process.

10.6. HIGH-T$_c$ MATERIALS

Most of the discussed superconducting materials have transition temperatures below 20 K. This is not very useful for applications at (or near) room temperature. However, a new era opened in the mid-1980s when superconductivity was discovered in $La_{2-x}Ba_xCuO_4$ at a temperature of about 30 K. The first reports of this new material caused immediate work in a number of laboratories, and new materials were found which raised the transition temperature into the 120–160 K range. While still not at room temperature, these discoveries reinvigorated the search for high temperature superconductors.

A common characteristic of all the new materials is that they are ceramics — often called flaky oxides, since they tend to be layered materials. These are poor metals at room temperature, which fits the discussed requirements. However, they do not fit the BCS theory and are highly anisotropic materials. Many of the materials are quite easy to make in polycrystalline form, but are exceedingly difficult to make in single crystal form. Following the intensive work that has occurred in the last 15 years, it is now believed that the behavior of carrier dynamics in the Cu–O planes is crucial to the superconductivity. We illustrate one of these materials in Fig. 10.10 — $YBa_2Cu_3O_7$, the so-called 1-2-3 compound (due to the relative number of atoms of the first three constituents). Note that there are distinct layers of this material. The top and bottom layer of atoms in each material layer (there are roughly three atomic layers in each so-called composite layer) are essentially Cu–O layers, as is the central layer. However, the atomic arrangement of the central layer is significantly different from that of the other two layers. The two outer layers (the top and bottom layers) tend to have the oxygen arranged in a hexagonal structure, with an oxygen atom at the center of each hexagon. Within this layer, two Cu atoms per hexagon appear. On the other hand, the central layer is formed with the oxygen arranged in rectangles, with

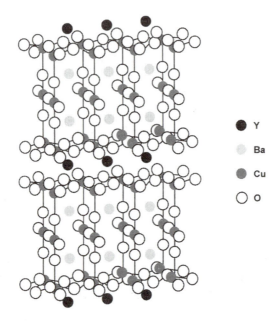

Figure 10.10. The crystal structure of YBCO ($YBa_2Cu_3O_{7-x}$).

a Cu atom at each end of the rectangle. The Ba atoms sit between
the three layers, while the Y atoms sit between each of the composite
layers (Fig. 10.10 shows two composite layers.) The conducting layers
are thought to be the top and bottom layers, while the central layer
serves as a charge reservoir.

A key issue is that the compound is antiferromagnetic—that is,
the magnetization of each of the composite layers is oppositely
directed. This means that there is a large magnetic moment in each
composite layer, but the sum over two layers vanishes. If one injects
holes into the central Cu–O planes, the system changes into a
superconductor. This can be accomplished by adding additional O
atoms. However, beyond this rudimentary understanding, there is *no*
theory for the appearance of superconductivity in these compounds.
As with many other topics discussed in this book, there are many
good ideas, but the absence of a predictive theory has eluded many
of the best minds in the business. In high-temperature superconductivity, as well as in other areas, there is room for new ideas that may
provide the crucial understanding.

PROBLEMS

1. Calculate the energy gap for lead as a function of temperature near T_c if $T_c = 7.2$ K, $E_{G0} = 2.7$ meV.
2. In the normal state, lead behaves as a nearly free-electron metal, with one free electron per atom and has a face-centered-cubic lattice of $a_0 = 0.494$ nm. Lead has a density of 11.34 g/cm³. Compute the free electron density and the London penetration depth.
3. Tin has the diamond lattice, with $a_s = 0.575$ nm, and has a density of 5.75 g/cm³. Compute the free-electron density and the London penetration depth.
4. From the results of Prob. 3, calculate $\rho(E_F)$ and E_F. Then, using the fact that $T_c = 3.72$ K and $T_D = 199$ K, calculate the interaction potential U.
5. The energy gap in lead is 2.7 meV. The Fermi energy level is 6.95 eV. Calculate the critical velocity v_c and the current density at this velocity.
6. In a Josephson junction, the maximum frequency which can be generated is the frequency corresponding to the energy gap, 2Δ. Calculate the maximum frequency from junctions of lead and tin. What are the wavelengths corresponding to these frequencies? Suppose that the junctions are irradiated at a frequency of 10^{10} Hz; what voltage is generated in the junction?

APPENDIX A

The Hydrogen Atom

In Sec. 3.8, we considered the atomic structure of the atoms on a very *ad hoc* basis. At that time, we pointed out that the original Bohr model of the atom was wrong in certain critical details. One of these is that the orbital angular momentum of the electron was quantized in units of $\sqrt{l(l+1)}/\hbar$ rather than $n\hbar$, as was assumed in the Bohr model. It is in the details of the quantization that the Bohr model fails. As we shall see here, the energy levels of electrons are not changed by a more proper quantum-mechanical treatment, a fact that must result from the theory, since the values of the energy produced from the Bohr model compare quite accurately with those obtained by experiment. In order to provide insight into the more rigorous treatment of the atom, in this appendix we shall consider the quantum-mechanical solution to the hydrogen atom.

If the nucleus of the hydrogen atom is considered as being fixed in position (we take it as the center of the coordinate system), the Schrödinger equation for the motion of the electron can be written as

$$-\frac{\hbar^2}{2m}\nabla^2\psi(\mathbf{r}) - \frac{e^2}{4\pi\varepsilon_0 r}\psi(\mathbf{r}) = E\psi(\mathbf{r}), \qquad (A.1)$$

where the second term on the left-hand side is the potential energy arising from the coulomb interaction between the electron and the

393

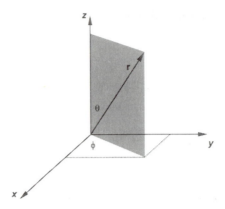

Figure A.1. Polar coordinate system, indicating the definition of the symbols $\mathbf{r}$, θ, and ϕ.

nucleus, when separated by a distance r. The first term on the left-hand side of (A.1) is the three-dimensional *Laplace operator*, defined as

$$
\begin{aligned}
\nabla^2 &= \frac{\partial^2}{\partial x^2} + \frac{\partial^2}{\partial y^2} + \frac{\partial^2}{\partial z^2} \\
&= \frac{1}{r^2} \frac{\partial}{\partial r} \left(r^2 \frac{\partial}{\partial r} \right) + \frac{1}{r^2 \sin^2\theta} \frac{\partial^2}{\partial \varphi^2} + \frac{1}{r^2 \sin\theta} \frac{\partial}{\partial \theta} \left(\sin\theta \frac{\partial}{\partial \theta} \right),
\end{aligned}
\tag{A.2}
$$

where the first line is in Cartesian coordinates. In the case of the one-electron atom, however, it is more practical to take advantage of the spherical symmetry of the problem and use spherical coordinates. In this case, the second line of (A.2) gives the Laplace operator in these spherical coordinates. The various angles are defined in Fig. A.1.

In Chapter 3, we split the Schrödinger equation into its spatial components and its temporal components by a process of separation of variables. It is possible to do the same thing here in the three coordinates. To do this, we write the wave function as a product of functions of each of the separate variables, which in spherical coordinates becomes

$$
\psi(\mathbf{r}) = R(r)\Theta(\theta)\Phi(\varphi).
\tag{A.3}
$$

When this form is inserted into the Schrödinger equation (A.1), using the second line of (A.2) for the Laplace operator, and the result is then divided by the wave function, we arrive at the form

$$\frac{1}{R}\frac{\partial}{\partial r}\left(r^2\frac{\partial R}{\partial r}\right) + \frac{2mr^2}{\hbar^2}\left(E + \frac{e^2}{4\pi\varepsilon_0 r}\right)$$

$$= -\left[\frac{1}{\Phi\sin^2\theta}\frac{\partial^2\Phi}{\partial\varphi^2} + \frac{1}{\Theta\sin\theta}\frac{\partial}{\partial\theta}\left(\sin\theta\frac{\partial\Theta}{\partial\theta}\right)\right]. \qquad (A.4)$$

The left-hand side of (A.4) is a function of r alone, while the right-hand side is a function of only θ and φ. If these are to be equal to each other for all values of r, φ, and θ, then each must be equal to a constant. This yields two equations, each of which is equal to the constant λ, as

$$\frac{1}{R}\frac{\partial}{\partial r}\left(r^2\frac{\partial R}{\partial r}\right) + \frac{2mr^2}{\hbar^2}\left(E + \frac{e^2}{4\pi\varepsilon_0 r}\right) = \lambda,$$

$$\left[\frac{1}{\Phi\sin^2\theta}\frac{\partial^2\Phi}{\partial\varphi^2} + \frac{1}{\Phi\sin\theta}\frac{\partial}{\partial\theta}\left(\sin\theta\frac{\partial\Theta}{\partial\theta}\right)\right] = -\lambda. \qquad (A.5)$$

We can now solve these two equations separately.

A.1. SEPARATION OF THE ANGULAR EQUATION

The second equation of (A.5) still contains both functions Θ and Φ. However, it can be further separated by rewriting it as

$$\frac{1}{\Phi}\frac{\partial^2\Phi}{\partial\varphi^2} = -\lambda\sin^2\theta - \frac{\sin\theta}{\Theta}\frac{\partial}{\partial\theta}\left(\sin\theta\frac{\partial\Theta}{\partial\theta}\right). \qquad (A.6)$$

The left-hand side of (A.6) is a function of φ alone, while the right-hand side is a function of θ alone. If this is to be valid for all values of φ and θ, then each side must be equal to a constant, which is taken as $-m_z^2$. The subscript is included to distinguish the constant from the mass m of the electron. This subscript is taken as z, since ϕ is rotation about the z axis, and it will be found that m_z is a result for the ϕ equation. Then we can rewrite (A.6) as

$$\frac{d^2\Phi}{d\varphi^2} + m_z^2\Phi = 0,$$

$$\frac{1}{\sin\theta}\frac{\partial}{\partial\theta}\left(\sin\theta\frac{\partial\Theta}{\partial\theta}\right) + \left(\lambda - \frac{m_z^2}{\sin^2\theta}\right)\Theta = 0. \tag{A.7}$$

From the first equation of (A.7), we observe that $\Phi(\phi)$ has a simple harmonic motion, and we can write the solution as

$$\Phi(\varphi) = A e^{im_z\varphi}, \tag{A.8}$$

with

$$m_z = 0, \pm 1, \pm 2, \ldots . \tag{A.9}$$

We expect that m_z will be limited in its range of integers, and this limitation will be found from the consideration of the second angular equation. To be completely valid, the wave function for ϕ must be normalized. We may readily accomplish this by requiring

$$\int_0^{2\pi} A^2 |e^{im_z\varphi}|^2 \, d\varphi = 2\pi A^2, \tag{A.10}$$

or

$$A = \frac{1}{\sqrt{2\pi}}. \tag{A.11}$$

To facilitate the handling of the second equation of (A.6), it is useful to introduce the change of variable

$$u = \cos\theta. \tag{A.12}$$

Then, we can rewrite this equation as

$$\frac{d}{du}\left[(1-u^2)\frac{d\Theta(u)}{du}\right] + \left[\lambda - \frac{m_z^2}{1-u^2}\right]\Theta(u) = 0. \qquad (A.13)$$

Expanding the first term yields the new form

$$(1-u^2)\frac{d^2\Theta(u)}{du^2} - 2u\frac{d\Theta(u)}{du} + \left[\lambda - \frac{m_z^2}{1-u^2}\right]\Theta(u) = 0. \quad (A.14)$$

While this does not seem to be any easier to solve, it turns out that this is a "well-known" equation. Those *thoroughly* familiar with mathematical boundary value problems will recognize (A.14) as the *associated Legendre equation*, whose solutions are the set of associated Legendre polynomials. However, we may readily solve for $\theta(u)$ by a series method, after a simple modification. The last term in the brackets is going to cause difficulty at $u = \pm 1$ ($\theta = \pm \pi$). Therefore, we define the form of $\Theta(u)$ as

$$\Theta(u) = (1-u^2)^{|m_z|/2}P(u), \qquad (A.15)$$

and we can rewrite (A.14) (taking m_z as the magnitude, and hence positive) as

$$(1-u^2)\frac{d^2P(u)}{du^2} - 2u(m_z+1)\frac{dP(u)}{du} + [\lambda - m_z - m_z^2]P(u) = 0. \quad (A.16)$$

To solve this, we assume a series solution of the form

$$P(u) = \sum_i a_i u^i. \qquad (A.17)$$

When this series is inserted into (A.16), we obtain the resultant

$$\sum_i i(i-1)a_i(u^{i-2} - u^i) - \sum_i 2(m_z+1)ia_i u^i + \sum_i [\lambda - m_z - m_z^2]a_i u^i = 0.$$

$$(A.18)$$

All terms are of the same order in i, except the first term. In this term, we replace i by $i + 2$, so that we can rewrite the above equation as

$$\sum_i \{(i+2)(i+1)a_{i+2} - [i(i-1) + 2i(m_z+1) + m_z(m_z+1) - \lambda]a_i\}u^i = 0.$$

(A.19)

Equation (A.19) is a *recursion relation*, which connects terms which differ by u^2. Thus there are two independent solutions, and these are the series for odd and even values of the exponent. The terms a_0 and a_1 are the two constants that are unknown and the other coefficients are related to them by the recursion formula

$$a_{i+2} = \frac{i(i+1) + 2i(m_z+1) + m_z(m_z+1) - \lambda}{(i+2)(i+1)} a_i.$$

(A.20)

At some maximum value of i, the series must terminate. If it did not, then it would not necessarily cause P to converge at $u = 1$. Since this function must converge, it is required that the series converge. This may be accomplished, for a maximum value of i taken to be l, which is related to λ, as

$$a_{l+2} = 0; \qquad l(l+2) + 2l(m_z+1) + m_z(m_z+1) = \lambda. \quad (A.21)$$

This result must hold for any value of m_z. To find this absolute value, we take $m_z = 0$, so that the required limit may be achieved by taking the value of l to be related to λ by

$$\lambda = l(l+1).$$

(A.22)

Then, the actual maximum value of the polynomial is i_{max}, for the case in which $m_z \neq 0$, may be found from (A.20) as

$$i_{max}(i_{max}+1) + 2i_{max}(m_z+1) + m_z(m_z+1) - l(l+1) = 0. \,(A.23)$$

This results in the actual limit to be

$$i_{max} = l - |m_z|.$$

(A.24)

Since this must be a positive integer, we also achieve the limiting value for the azimuthal quantum number

$$|m_z| \leqslant l. \tag{A.25}$$

Combining these results, we achieve the final value for the polar wave function to be

$$\Theta(u) = P_l^{m_z}(u) = (1 - u^2)^{|m_z|/2} \sum_{i=0,1}^{l-|m_z|} a_i u^i, \tag{A.26}$$

where the summation is taken *only* over the odd or even values of i, depending on whether $l - |m_z|$ is odd or even. The first few normalized (in the interval $-1 \leqslant u \leqslant 1$) associated Legendre polynomials are

$$P_0^0(u) = \frac{1}{\sqrt{2}}, \qquad\qquad \Theta_0^0(\theta) = \frac{1}{\sqrt{2}},$$

$$P_1^0(u) = \sqrt{\frac{3}{2}}\, u, \qquad\qquad \Theta_1^0(\theta) = \sqrt{\frac{3}{2}}\cos\theta,$$

$$P_1^1(u) = \sqrt{\frac{3}{4}}\sqrt{1 - u^2}, \qquad \Theta_1^1(\theta) = \sqrt{\frac{3}{4}}\sin\theta,$$

$$P_2^0(u) = \sqrt{\frac{5}{8}}(3u^2 - 1), \qquad \Theta_2^0(\theta) = \sqrt{\frac{5}{8}}(3\cos^2\theta - 1),$$

$$P_2^1(u) = \sqrt{\frac{15}{4}}\, u\sqrt{1 - u^2}, \qquad \Theta_2^1(\theta) = \sqrt{\frac{15}{4}}\sin\theta\cos\theta,$$

$$P_2^2(u) = \sqrt{\frac{15}{16}}(1 - u^2), \qquad \Theta_1^1(\theta) = \sqrt{\frac{15}{16}}\sin^2\theta.$$

From the separation of the angular equation, we have found the values of two of the separation constants. These are $\lambda = l(l + 1)$, and $m_z = -l, -l+1, \ldots -1, 0, 1, l - 1, l$. Here, m_z is an integer, which means that l is also an integer, which is required by the evaluations of the series above.

A.2. THE RADIAL EQUATION

When we use the value of the separation constant λ found in the previous section, the radial equation (first equation) of (A.5) becomes

$$\frac{1}{r^2}\frac{\partial}{\partial r}\left(r^2\frac{\partial R}{\partial r}\right) + \left\{\frac{2m}{\hbar^2}\left(E + \frac{e^2}{4\pi\varepsilon_0 r}\right) - \frac{l(l+1)}{r^2}\right\}R = 0. \quad (A.27)$$

To remove some of the extraneous constants, let $\rho = \alpha r$. At the same time, it is desirable to replace the energy E with a numerical constant, which will be found after the value of α is properly chosen. If we insert $\rho = \alpha r$ into (A.27), the result is

$$\frac{1}{\rho^2}\frac{\partial}{\partial \rho}\left(\rho^2\frac{\partial R}{\partial \rho}\right) + \left\{\frac{2m}{\hbar^2\alpha^2}\left(E + \frac{e^2\alpha}{4\pi\varepsilon_0 \rho}\right) - \frac{l(l+1)}{\rho^2}\right\}R = 0, \quad (A.28)$$

which can be rewritten in the form

$$\frac{1}{\rho^2}\frac{\partial}{\partial \rho}\left(\rho^2\frac{\partial R}{\partial \rho}\right) + \left\{\frac{2m}{\hbar^2\alpha^2}\frac{e^2\alpha}{4\pi\varepsilon_0 \rho} + \frac{2mE}{\hbar^2\alpha^2} - \frac{l(l+1)}{\rho^2}\right\}R = 0. \quad (A.29)$$

Looking ahead, we want to have the second term in the curly brackets to take the value $1/4$, which means that we must define α as

$$\alpha^2 = \frac{8m|E|}{\hbar^2}. \quad (A.30)$$

With this definition, the coefficient of the $1/\rho$ term, the first term in the curly brackets, becomes

$$Q = \frac{me^2}{2\pi\varepsilon_0\hbar^2\alpha} = \frac{e^2}{4\pi\varepsilon_0\hbar}\sqrt{\frac{m}{2|E|}}. \quad (A.31)$$

Recognizing that, for a bound state, the energy is negative, it is apparent why the magnitude was introduced into these two equations. Then, the radial equation becomes

$$\frac{1}{\rho^2}\frac{\partial}{\partial\rho}\left(\rho^2\frac{\partial R}{\partial\rho}\right) + \left\{\frac{Q}{\rho} - \frac{1}{4^2} - \frac{l(l+1)}{\rho^2}\right\}R = 0. \qquad (A.32)$$

Equation (A.32) is second order in the normalized radial distance ρ. For large values of ρ, we can easily see that the solutions are exponentially decaying waves, varying as $e^{-\rho/2}$. Hence, just as in the last section, we will make another transformation by assuming that

$$R(\rho) = F(\rho)e^{-\rho/2}. \qquad (A.33)$$

With this change, we can seek a power series solution for $F(r)$. Inserting (A.33) into (A.32) gives rise to the new equation:

$$\frac{d^2F}{d\rho^2} + \left(\frac{2}{\rho} - 1\right)\frac{dF}{d\rho} + \left[\frac{Q-1}{\rho} - \frac{l(l+1)}{\rho^2}\right]F = 0. \qquad (A.34)$$

To solve this equation, we first use the substitution and power series

$$F(\rho) = \rho^s L(\rho) = \rho^s \sum_j a_j \rho^j. \qquad (A.35)$$

This leads to

$$\rho^2\frac{d^2L}{d\rho^2} + \rho[2(s+1)-\rho]\frac{dL}{d\rho} + [\rho(Q-1-s)+s(s+1)-l(l+1)]L = 0. \qquad (A.36)$$

This equation must be valid for every value of ρ. In particular, at $\rho = 0$, we have

$$[s(s+1) - l(l+1)]L = 0. \qquad (A.37)$$

Since, in general, we do not require $L = 0$, we must then have

$$s = l, -(l+1). \qquad (A.38)$$

These two solutions give the two distinct solutions to the radial equation. However, for $F(\rho)$ to be finite at $\rho = 0$, we must discard the second solution by choosing its coefficient to be zero under all circumstances. Hence, we are left with the single value $s = l$. Then, (A.36) becomes

$$\rho\frac{d^2L}{d\rho^2} + [2(l+1) - \rho]\frac{dL}{d\rho} + \rho(Q - 1 - l)L = 0. \qquad (A.39)$$

We now insert the power series for L into this equation, which yields

$$\sum_j \{a_j j(j-1)\rho^{j-1} + 2(l+1)ja_j\rho^{j-1} - ja_j\rho^j + (Q-1-l)a_j\rho^j\} = 0.$$

$$(A.40)$$

We can change the indices on the first two terms, so that all the powers of ρ are the same, and this leads to

$$\sum_j \{a_{j+1}j(j+1) + 2(l+1)(j+1)a_{j+1} - ja_j + (Q-1-l)a_j\}\rho^j = 0.$$

$$(A.41)$$

This equation yields the recursion relation on the coefficients to be

$$a_{j+1} = a_j\frac{j - (Q - 1 - l)}{(j + 2l + 2)(j + 1)}. \qquad (A.42)$$

This series has a problem, in that for large j, the ratio

$$\frac{a_{j+1}}{a_j} \rightarrow \frac{1}{j}, \qquad (A.43)$$

which means that this series diverges faster than an exponential. Hence, the solution for $R(\rho)$ will diverge for large ρ. The way around this problem is to employ the same requirement used in the last section. That is, we require that the series not be an infinite series, but must terminate at some power j_{max}. Hence, the numerator of

(A.42) must vanish at this power, or

$$j_{max} + l + 1 - Q = 0. \tag{A.44}$$

An immediate consequence of this limitation is that Q *must be an integer*! That is,

$$Q = n, \qquad n = 1, 2, 3, \ldots,$$

$$E = -\frac{m}{2}\left(\frac{e^2}{4\pi\varepsilon_0 \hbar Q}\right)^2 = -\frac{me^4}{32\pi^2\varepsilon_0^2\hbar^2 n^2}. \tag{A.45}$$

Finally, the radial wave function is then

$$R_{nl}(r) = (\alpha r)^l e^{-\alpha r/2} \sum_{j=0}^{n-(l+1)} a_j(\alpha r)^j \tag{A.46}$$

While the energy depends only upon the radial quantum number n, the wave function depends both upon the radial quantum number and the polar quantum number l. It can be shown that this polar quantum number actually is a result of quantizing the angular momentum L (which should not be confused with the preceding series), with the result that

$$L^2 = l(l + 1)\hbar^2. \tag{A.47}$$

This quantization of the angular momentum is an important result of the quantum theory of the atom.

APPENDIX B

Impurity Insertion

We now want to turn our attention to the processing details required to actually put impurities into a semiconductor in controlled locations. The oldest method, of course, of intentionally introducing impurities was to include them in the melt during the growth of the bulk semiconductor. With the advent of planar processing over entire wafers, sliced from the bulk, this became no longer a desired method of controlling the impurity concentration at specific sites on the wafer. Instead, the diffusion of impurities, through an oxide mask, was adopted for this purpose. Later, ion implantation became a method of choice. In this appendix, we discuss the details of the spatial distribution of dopants following diffusion and ion implantation.

B.1. DIFFUSION OF IMPURITIES

In Fig. B.1, we indicate the method by which impurities are diffused into a semiconductor wafer at high temperatures. An oxide layer is grown on the semiconductor, and then openings in this oxide are etched by a lithographic process. The openings exist at the sites on the wafer where impurity diffusion is desired. The wafer is then placed in an oven at high temperature in a gas containing the impurities.

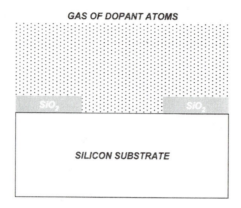

Figure B.1. Diffusion of dopant atoms into a silicon substrate. The SiO$_2$ layer is used as a mask to ensure that diffusion into the substrate is limited to a lithographically defined region.

However, there is a problem in that the silicon wafer will tend to grow an oxide in the opened hole under most conditions. Thus, there are two different diffusion conditions. In one case, the opening remains open during the entire diffusion process, for which the impurity profile will be the integral of the Gaussian solution of the diffusion equation, or

$$N(x, t) = N_0 \operatorname{erfc}\left(\sqrt{\frac{x^2}{4Dt}}\right), \qquad (B.1)$$

where N_0 is the surface concentration determined by the gas itself, D is the diffusion constant, t is the diffusion time, and x is the direction normal to the surface. The special function is the *complementary error function.*

On the other hand, one often deposits the impurity at very high temperature, then allows the oxide to form. Diffusion is then carried out in a subsequent temperature cycle. The predeposition of the impurities in this situation means that a fixed number of total impurites will exist in the semiconductor, and the distribution is

$$N(x, t) = \frac{N_{S,\text{Total}}}{\sqrt{\pi Dt}} \exp\left(-\frac{x^2}{4Dt}\right), \qquad (B.2)$$

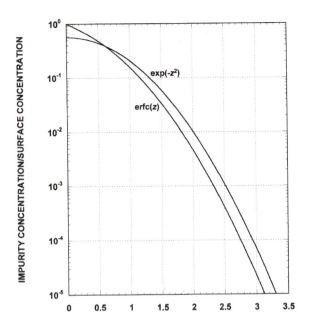

Figure B.2. Normalized impurity concentration versus normalized distance for error-function-complement and Gaussian distributions. These are normalized to give the same number of impurities per unit surface area.

where $N_{S,\text{Total}}$ is the sheet concentration of the impurities in the predeposition process (number per square meter). The difference in these two curves for the same total number of impurities is shown in Fig. B.2.

The diffusion constant D is itself temperature dependent, as diffusion is an activated process by which the impurities move through the lattice. This movement can be either by the impurity sitting in an interstitial site (between the lattice atoms) or by displacing a lattice atom. The diffusion constant is typically given as

$$D(N, T) = D_0(N) \exp\left(-\frac{E_A}{k_B T}\right), \qquad (B.3)$$

where E_A is the activation energy, and $D_0(N)$ is the "infinite temperature" value of the diffusion constant. The dependence on the concen-

TABLE B.1. Dopant Diffusion in
Silicon

	D_0	E_A
B	$30\,\text{cm}^2/\text{s}$	$3.85\,\text{eV}$
As	5.34	3.87
$P_1{}^a$	3.32	2.65
$P_2{}^a$	1.6	2.65

[a]The first value is for a background doping of 10^{17} while the second value is for
$5 \times 10^{14}\,\text{cm}^{-3}$.

tration N is used to indicate that many impurities, such as B, have a
concentration dependence in their diffusion properties. In Table B.1,
we list the parameters for some common dopants in silicon.

B.2. ION IMPLANTATION

In ion implantation, a similar procedure to that just described is
followed, in that a mask (typically an oxide) is used. A hole is opened
where the impurities are desired. However, here the similarity ends.
Instead of putting the wafer into a gas of the impurity, the impurity
is ionized and then accelerated to high voltage. A beam of the ionized
impurities is directed at the wafer and will penetrate into the wafer
where the mask has been opened. The impurities lose energy to the
atomic electrons as well as the vibrations of the lattice atoms. As they
lose energy, they slow down, and eventually stop at a depth below
the surface that is determined by their initial energy. This depth is
called the *range* of the implant. Because the beam is a statistical entity
and the scattering is a random process, the implanted impurities have
a distribution about the range according to

$$N(x) = \bar{N} \exp\left(-\frac{(x - R_p)^2}{(\Delta R_p)^2}\right). \tag{B.4}$$

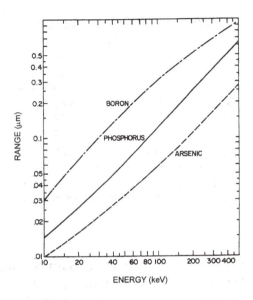

ENERGY (keV)

Figure B.3. Projected range versus energy for boron, phosphorous, and arsenic ions in silicon. *Reproduced with permission from "Ion implantation in silicon—physics, processing, and micro-electronic devices," by K. A. Pickar, in Applied Solid State Science, vol. 5, Ed. R. Wolfe (Academic Press, 1975).*

Here, R_p is the range of the implant, $\bar{N}$ is the peak concentration, and ΔR_p is called the *straggle* and describes the statistical spread of the impurities around the peak at the depth given by the range. Since the impurities can significantly damage the lattice during this process, the wafer is then annealed, typically to greater than 900°C, to restore the lattice integrity. At the same time, the impurities can move onto the lattice sites and become electrically active. In Figs. B.3 and B.4, we plot the experimentally determined range and straggle, respectively, for common dopants in silicon. One important point is that the range in SiO_2 is quite similar to that in Si, so that it is important that the oxide mask be considerably thicker than the projected range of the implant if it is to be an effective mask.

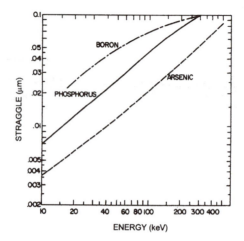

Figure B.4. Straggle versus energy for boron, phosphorous, and arsenic ions in silicon. *Reproduced with permission from "Ion implantation in silicon — physics, processing, and micro-electronic devices," by K. A. Pickar, in Applied Solid State Science, vol. 5, Ed. R. Wolfe (Academic Press, 1975).*

APPENDIX C

Semiconductor Properties

Parameter		Si	Ge	GaAs
Band gap	E_G (eV) (300 K)	1.1	0.66	1.4
Number of equivalent minima in CB		6	4	1
Effective mass, electron	m_e/m_0	0.32	0.22	0.07
Effective mass, hole	m_h/m_0	0.6	0.37	0.5
Effective density of states	N_c (cm^{-3})	2.72×10^{19}	1.03×10^{19}	4.34×10^{17}
Effective density of states	N_v (cm^{-3})	1.16×10^{19}	3.93×10^{18}	8.85×10^{18}
Electron mobility	μ_e (cm^2/Vs)	1500	3800	7500
Hole mobility	μ_h (cm^2/Vs)	500	2000	450
Electron diffusivity	D_e (cm^2/s)	38.9	98.4	194
Hole diffusivity	D_h (cm^2/s)	13.0	51.8	11.7
Intrinsic concentration	n_i (cm^{-3})	1.0×10^{10}	8.4×10^{12}	3.5×10^6
Dielectric constant, static	ε_{r0}	11.9	15.9	11.1
Dielectric constant, optical	$\varepsilon_{r\infty}$	11.9	15.9	13.1
Density	ρ (g/cm^3)	2.33	5.33	5.35

APPENDIX D

Some Fundamental Constants

Electron charge	e	$1.6 \times 10^{-19}\,\mathrm{C}$
Electron mass	m_0	$9.1 \times 10^{-31}\,\mathrm{kg}$
Permittivity of free space	ε_0	$8.854 \times 10^{-12}\,\mathrm{F/m}$
Speed of light in vacuum	c	$3.0 \times 10^{8}\,\mathrm{m/s}$
Boltzmann's constant	k_B	$1.38 \times 10^{-23}\,\mathrm{J/K}$
Planck's constant	h	$6.626 \times 10^{-34}\,\mathrm{J \cdot s}$
Planck's reduced constant	$\hbar = h/2\pi$	$1.05459 \times 10^{-34}\,\mathrm{J \cdot s}$
Avogadro's number	N_A	$6.025 \times 10^{23}/\mathrm{mole}$
Bohr magneton	β	$9.27 \times 10^{-24}\,\mathrm{A\ m^2}$
Bohr radius in atom	a_0	$5.3 \times 10^{-11}\,\mathrm{m}$
Rydberg energy	E_R	$13.5\,\mathrm{eV}$

INDEX

Index